Evaluating Philosophies

BOSTON STUDIES IN THE PHILOSOPHY OF SCIENCE

VOLUME 295

For further volumes:
http://www.springer.com/series/5710

Mario Bunge

Evaluating Philosophies

Springer

Mario Bunge
Department of Philosophy
McGill University
Montreal, Canada

ISSN 0068-0346
ISBN 978-94-007-4407-3 ISBN 978-94-007-4408-0 (eBook)
DOI 10.1007/978-94-007-4408-0
Springer Dordrecht Heidelberg New York London

Library of Congress Control Number: 2012941366

Printed on acid-free paper

Springer is part of Springer Science+Business Media (www.springer.com)

Preface

This book tries to answer a question that laymen never cease to ask: How can the worth of a philosophy be assessed? The work consists of a complaint, a proposal, and answers to two groups of neglected philosophical questions. The complaint refers to the doctrines that block the advancement of knowledge even though they are passed off as philosophies. And the proposal is to measure philosophies by the way they help inquire or act.

The bulk of this volume is taken by a sample of problems that have been either neglected or approached in a dubious manner. While some of these problems touch on social praxis, others occur in the course of scientific or humanistic research. There is also a short glossary.

I thank Marcelo H. Bosch, Carlos F. Bunge, Silvia A. Bunge, Antoni Domenech, Bernardo Gabarain, Rafael González del Solar, Irving Louis Horowitz, Víctor Landman, Javier López de Casenave, Antonio Martino, José Luis Pardos, Nicholas Rescher, Gustavo E. Romero, Dan A. Seni, and Héctor Vucetich for valuable exchanges.

Contents

Part I How to Nurture or Hinder Research

1 Philosophies and Phobosophies 3
- 1.1 Midwives 4
- 1.2 Teachers 6
- 1.3 Gatekeepers 7
- 1.4 Wardens and Prisoners 7
- 1.5 Cheated 9
- 1.6 Mercenary 10
- 1.7 Escapist 11
- 1.8 Ambivalent 12
- 1.9 Conclusion 13

2 The Philosophical Matrix of Scientific Progress 15
- 2.1 From Skepticism to Mysterianism 15
- 2.2 The Social Matrix 16
- 2.3 The Role of Philosophy in the Birth of Modern Science 17
- 2.4 Materialism, Systemism, Dynamicism, and Realism 20
- 2.5 First Parenthesis: The Ossification of Philosophy 22
- 2.6 Scientism, Rationalism, and Humanism 24
- 2.7 Second Parenthesis: Logical Imperialism 27
- 2.8 The Philosophical Pentagon 28
- 2.9 Irregular Pentagons 29
- 2.10 From Social Science to Sociotechnology 30
- 2.11 Dogmatic and Programmatic *Isms* 30
- 2.12 Concluding Remarks 32
- References 33

3 Systemics and Materialism 35
- 3.1 The Housing Problem: A Component of a Ten-Dimensional Problem 35
- 3.2 Approach 37

3.3 Preliminary Examples 37
3.4 Systemic Approach and Theory 39
3.5 Natural Sciences 40
3.6 Social Sciences 41
3.7 Biosocial Sciences 42
3.8 Technologies 43
3.9 The Knowledge System 44
3.10 Philosophical Systems 45
3.11 Concluding Remarks 46
References 46

Part II Philosophy in Action

4 Technoscience? 51
4.1 Discovery and Invention 51
4.2 Primacy of Praxis? 52
4.3 Consequences of the Confusión 53
4.4 "Translation" of Science into Industry via Technology 54
4.5 Authentic Technosciences 54
4.6 Conclusion 55
References 56

5 Climate and Logic 57
5.1 The Kaya Identity 57
5.2 From Logic to Reality 58
5.3 A New Formula 59
5.4 Conclusion 59
References 59

6 Informatics: One or Multiple? 61
6.1 From Information System to Communication System 61
6.2 Back to Information 63
6.3 Conclusion 64
References 64

7 Wealth and Well-being, Economic Growth and Integral Development 65
7.1 Is Happiness for Sale? 65
7.2 Can Well-Being Be Bought? 68
7.3 The Problem of Inequality 69
7.4 Sectoral Growth and Integral Development 71
7.5 Conclusions 73
References 75

8 Can Standard Economic Theory Account for Crises? 77
8.1 Standard Economics Focuses on Equilibrium 77
8.2 The Economic Rationality Postulate 79

8.3 The Free Market Postulate 80
8.4 Conclusion 81
References 82

9 Marxist Philosophy: Promise and Reality 83
9.1 Dialectical Materialism 84
9.2 Hegel's Disastrous Legacy 85
9.3 Historical Materialism 86
9.4 Epistemology and the Sociology of Knowledge 88
9.5 Theory and Praxis, Apriorism and Pragmatism 90
9.6 State and Planning 91
9.7 Dictatorship and Disaster 92
9.8 Conclusion 92
References 93

10 Rules of Law: Just and Unjust 95
10.1 Politics, Law, and Morals 95
10.2 Legal Legitimacy 96
10.3 Political Legitimacy 96
10.4 Moral Legitimacy and Legitimacy Tout Court 97
10.5 Emergencies 98
10.6 If You Wish Order, Prepare for Disorder 98
10.7 The Ultimate Test: The Rise of Nazism 99
10.8 Legal Positivism: Fig Leaf of Authoritarianism 101
10.9 Conclusion 102
References 102

Part III Philosophical Gaps

11 Subjective Probabilities: Admissible in Science? 105
11.1 Beware Ordinary Language 105
11.2 Monty Hall Brain Teasers 107
11.3 Learned Ignorance 108
11.4 Probability as Credence 109
11.5 Prior Probabilities Are Inscrutable 110
11.6 Bayesianism Can Be Disastrous 110
11.7 Trial by Numbers 111
11.8 Concluding Remarks 112
References 112

12 Does Inductive Logic Work? 115
12.1 The Semantic Objection to Inductivism 115
12.2 Bayesian Induction 116
12.3 A Radical Criticism of Bayesian Induction 117
12.4 Concluding Remarks 118
References 118

13 Bridging Theories to Data ... 119
13.1 The Attempt to Replace the Theoretical with the Empirical ... 119
13.2 Philosophers Rediscover Experiment ... 121
13.3 What Measuring Instruments Show ... 122
13.4 Visualizing the Unseen ... 123
13.5 Theory-Based Indicators ... 124
13.6 The Place of Indicators in Theory Testing ... 126
13.7 Indicators in Quantum Physics ... 127
13.8 Concluding Remarks ... 128
References ... 129

14 Matter and Energy: Physical or Metaphysical Concepts? ... 131
14.1 Energy: Stuff or Property? ... 131
14.2 Special Energies ... 132
14.3 Minitheory ... 133
14.4 Common Errors ... 136
14.5 The Ontological Square ... 137
14.6 Conclusion ... 137

15 Does Quantum Physics Refute Realism, Materialism and Determinism? ... 139
15.1 The Spiritualist View ... 139
15.2 Resorting to Experiment ... 140
15.3 Criticism of the Subjectivist Interpretation ... 141
15.4 Indeterminacy and Decoherence ... 142
15.5 The Quantum Theory Is Not About Observers ... 143
15.6 Quantum Theory and Realism: An Entangled Affair ... 144
15.7 Has Matter Vanished? ... 147
15.8 Indeterminism ... 147
15.9 Is Rationality Limited? ... 148
15.10 Conclusion ... 148
References ... 149

16 Parallel Universes? Digital Physics? ... 151
16.1 Imagination: Free and Disciplined ... 151
16.2 Possible Worlds ... 152
16.3 Virtual Worlds? Virtual Particles? ... 153
16.4 Digital Physics ... 154
16.5 Physics ≠ Semiotics ... 155
16.6 Concluding Remarks ... 156
References ... 156

17 Can Functionalist Psychology Explain Anything? ... 159
17.1 Functionalist Description ... 159
17.2 To Explain Is to Unveil Mechanisms ... 160
17.3 Submission Sickens ... 162
17.4 Beyond Nativism and Empiricism ... 163
17.5 Concluding Remarks ... 164
References ... 165

18 Knowledge Pyramids or Rosettes? 167
18.1 Epistemological Pyramids and Rosettes 167
18.2 Logical Pyramids 168
18.3 Ontological Pyramids 169
18.4 Place of the Universal Sciences 169
18.5 Concluding Remarks 170
References 171

19 Existence: Single or Double? 173
19.1 Existence and Someness 173
19.2 Existential Predicate and Anselm's Proof 174
19.3 Logic and Ontology 175
19.4 Truth: Single or Double? 176
19.5 Conclusion 178
References 178

20 Conclusion: Valuation Criterion 181
20.1 Monism or Pluralism? 181
20.2 The Fertility Criterion 182
20.3 Conclusion 183
References 183

Philosophical Glossary 185

Index 191

Introduction

How are philosophies evaluated? This question does not seem to have been investigated in any detail or depth. In any event, there do not seem to be any objective and generally accepted criteria for assessing the merits and flaw of philosophical doctrines. Usually, the adoption of a philosophy does not result from a long and anguished deliberation but, rather, from a combination of predisposition with necessity and opportunity – just as in the case of petty theft.

By contrast, to evaluate a science or a scientific theory scientists use a battery of objective criteria accepted by almost all investigators: clarity, internal consistency, fitness to the relevant empirical data, coherence with the bulk of antecedent knowledge, size of the problems it tackles, ability to answer extant questions, and potential to guide future research. Scientists and philosophers use these criteria and a few others every time the credentials of a new discipline or a theory are questioned. Suffice it to recall the scientific-philosophical controversies ignited by all the scientific breakthroughs during the past five centuries.

Nothing like that happens in philosophy: usually philosophical doctrines are accepted or rejected, fully or in part, without resorting to any clear and objective criteria. The evaluation of philosophical doctrines tends to be intuitive, utilitarian, or even emotive. For example, idealism was the philosophy of the Western establishment during the nineteenth century because it was part of the Counter-Enlightenment. Marx and his followers admired Hegel's dialectics, despite being hermetic and lacking in empirical support, just because they believed that, as Lenin claimed, it was "the algebra of revolution." The builders of modern Brazil adopted Comte's slogan, *Order and progress*, for it encapsulated the ideal of the enlightened planters.

Neo-Thomism was an attempt to rejuvenate the erstwhile official philosophy of the Catholic Church. In mid-twentieth century, the church considered replacing Thomism with phenomenology, but backtracked because Husserl's egology replaces God with the self. Mussolini embraced pragmatism because he regarded success as the supreme value; but he did not make it the official philosophy because, under a dictatorship, success is the preserve of the few. Heidegger served Nazism but his Party did no return the favor because existentialism was far too whining and hermetic. Linguistic philosophy attracts people who like clarity but avoid trouble and commitment. And at all times pseudoprofound aphorisms have been more popular than rigorous arguments.

In sum, philosophies, whether genuine or spurious, are not usually adopted because of their conceptual, empirical, or moral merits, but because of tradition, political interests, or even temperament – none of which is a good reason. The present book argues for a precise criterion: *A philosophy is worth what it helps learn, act, conserve our common heritage, and get along with fellow humans.*

Part I
How to Nurture or Hinder Research

Chapter 1
Philosophies and Phobosophies

A philosophical doctrine may either facilitate or hinder the exploration of reality. For example, realism favors such exploration, whereas constructivism-relativism is hostile to it. And a philosophy may be ambivalent: enlightened in some respects and obscurantist in others. For instance, Descartes' mind-body dualism was a bane for psychology and psychiatry because it encouraged brainless speculation , but a boon to biomedical research because it justified dissection and vivisection. Rousseau was progressive in political philosophy but reactionary with regard to science. Romanticism was revolutionary in art but reactionary in philosophy. Positivism exalted scientific research while hindering it for attempting to restrict it to describing appearances. And Marxism benefited the social sciences in stressing the centrality of the so-called material factors, but harmed it in underestimating other factors and in adopting Hegel's obscurities and confusions.

Besides, a philosophy that starts as progressive may become conservative upon being adopted by the establishment, as happened to Aristotelianism, Cartesianism, Leibnizianism, and Marxism. To help advance science and technology, a philosophy has got to keep pace with them instead of becoming ossified.

The preceding suggests that philosophers should pay far more attention to the sciences and technologies, and that scientists and technologists should evaluate philosophies by their fruits. For example, instead of treating phenomenology and linguistic philosophy in accordance with the principle of charity, they should ask whether those doctrines have helped knowing anything or whether, on the contrary, they have hindered the advancement of knowledge.

In what follows we shall glimpse at some of the major philosophies in their relation to the search for truth, in order to assess the proposal of measuring them according to the pragmatic criterion "By their fruits ye shall know them."

M. Bunge, *Evaluating Philosophies*, Boston Studies in the Philosophy of Science 295,
DOI 10.1007/978-94-007-4408-0_1, © Springer Science+Business Media Dordrecht 2012

1.1 Midwives

Every serious student, in any field, tries to reason correctly, that is, clearly and consistently. This is why the rationalist philosophies favor research. By contrast, the verbal acrobatics of Heidegger, Sartre, Derrida, Deleuze, Vattimo, Kristeva, Irigaray, and their imitators are either absurd or trivial.

I submit that postmodernism, in particular existentialism, are either absurd or trivial. For example, in his celebrated *Crisis of the European Sciences* (Part IIIA, §54b), Edmund Husserl wrote: "As primal ego, I constitute my horizon of transcendental others as cosubjects within the transcendental intersubjectivity which constitutes the world." In case of doubt, check the next page: "the immediate "I", already enduring in the enduring primordial sphere, constitutes in itself another as other. Self-temporalization through depresentation, so to speak (through recollection), has its an analogue in my self-alienation as a depresentation of a higher level–depresentation of my primal presence into a merely presententified primal presence". As Goya warned, "the slumber of reason produces monsters."

In any case postmodernism, in particular existentialism and "weak thinking", are against the advancement of knowledge because they simulate thought and lead students to memorizing formulas that they take to be profound just because they do not understand them. Their teachers do not dare telling them that the king is naked.

Without a modicum of rationality we do not get anywhere. Now, when the object under study is an ideal object, such as a number system or a theory, reason is both necessary and sufficient. But when the object of interest is concrete (material), radical rationalism like Leibniz's, is unsuitable, for it amounts to apriorism, which is just as arbitrary as irrationalism.

Nor is radical empiricism adequate to the sciences and technologies, because it underestimates theorizing about things, properties, and processes that, though real, are imperceptible, such as electrons and social relations. When attempting to know or to control something real, the suitable strategy is ratioempiricism, that is, a synthesis of rationalism and empiricism, which favors combining imagination with observation, and calculation with experiment.

However, not every merger of rationalism and empiricism will do. For example, Kant combined bad rationalism, namely apriorism, with bad empiricism, namely phenomenalism (sticking to appearances). By contrast, scientific realism combines the empiricist demand for empirical testing with the rationalist striving for hypotheses and theories accounting for appearances instead of sticking to them.

In sum, the ratioempiricist synthesis most favorable to the exploration of reality is the one that adopts the scientific approach. Paraphrasing what Engels said about Hegel, we claim that what the sciences and technologies must take from rationalism and empiricism are their methods (hypothetical-deductive reasoning and empirical testing respectively), not their systems–idealist ontology and phenomenalism respectively.

Now, if we wish to explore reality, we cannot limit ourselves to rejecting doctrines, the way Sextus Empiricus and other radial skeptics did: We must also hazard assumptions about what we want to know and can get to know. These assumptions are of two kind: ontological (about reality) and methodological (about research).

For example, the researcher may suppose that the universe is a continuum, or that it is constituted by particles; and he may assume that it is knowable either fully or to some extent, or that it is basically mysterious.

From the start of modernity at about 1,600, scientists and technologists have tended to conceive of things as systems or components of such, and to assume that they can be known bit by bit. In other words, scientists and technologists have increasingly embraced the systemic approach. This approach assumes that all things are either systems or components of such. Suffice it to recall the discoveries of the planetary and stellar systems, as well as those of the cardiovascular and nervous systems and, more recently, those of the neuro-endocrine-immune system, and the ecosystems and social systems, from the family to the international system.

The systemic hypothesis invites us to analyzing wholes as well as to disclose or neutralize the entities that may interact with them. Now, there are systems of various kinds: conceptual like theories and classifications, material such as molecules and organisms; and semiotics like texts and diagrams.

There is a tacit consensus that natural systems are material, and therefore cannot be influenced by pure ideas. For instance, we trust that our measuring instruments do not react to the wishes of their users. It is not that naturalists deny the mental: they only deny that it may exist outside some brain.

True, in social studies there are many who believe that everything social is spiritual–whence the name *Geisteswissenschaften*, or sciences of the spirit. This hypothesis leads to focusing on the symbolic aspects society: language, myths, norms, rituals, etc. But it is also true that the good contemporary anthropologist starts by finding out how the members of her favorite tribe make a living: *primum vivere*. The cave paintings may or may not have had a practical use, but there is no doubt that their authors were flesh and blood humans who performed manual tasks to get food, shelter, companionship, and security.

In sum, very scientist, whether natural or social, employs more or less tacitly a bundle of philosophical principles, among which the following stand out: clarity, rationality, realism, empirical testing, and systemism. The philosophies that contain these principles can perform the midwifery function that Socrates assigned himself.

There is little doubt that the pre-Socratics prepared the ground for the birth and growth of ancient science, in sketching a naturalist worldview, in particular the atomic theory. It is also true that, at about the same time, India produced a similar conception–but did not engender science. This suggests that naturalism (or materialism), though necessary to produce science, is not sufficient. Curiosity and intellectual courage, too, are required, and those were traits of an elite of thinkers and traders in ancient Greece.

Besides, as Carmen Dragonetti has emphasized, the Hellenes lacked a corpus of allegedly sacred scriptures that, like the *Veda*, claimed to explain everything. Nor did they have a priestly caste, like that of the Brahmins, in charge of spreading that text and see to it that its prescriptions be observed. Whereas the Indian intellectuals were subjected to that caste and kept bound to magic-religious thought, the original Greek thinkers took the liberty of questioning dogmas and demanding proof or evidence: they were philosophers in the original sense of this word.

1.2 Teachers

Philosophers can help scientists analyze and refine their concepts, as well as bring their presuppositions up to light. They can also help them argue better and question hypotheses, methods and findings that look obvious for being old or for having been endorsed by great scientists. Philosophical criticism may help progress in identifying obstacles to it. Let us recall a few recent examples of such weeding.

String theory is still fashionable in physics despite having failed to deliver anything plausible in the course of four decades. The philosopher of science is justified in suspecting that this theory is pseudoscientific because it violates a basic principle, first stated eight decades ago by Bohr and Einstein: that a new theory should not give up the firm acquisitions of its predecessors. (True, the enthusiasts of scientific revolutions, such as Bachelard, Kuhn, and Feyerabend, overlooked this principle–at their risk).

Clearly, string theory violates that principle, in postulating that space-time has 11 dimensions rather than 4: this assumption contradicts the rest of physics, which renders it implausible. Similarly, digital physics, in postulating that the basic constituents of the universe are *bits*, and thus symbols instead of physical entities, breaches the definition that everything physical has only physical properties, such as energy.

Another case in point is the Copenhagen interpretation of the quantum theory, which holds that all its formulas refer to experimental situations. Even a cursory examination of the basic formulas of this theory fails to reveal such reference to experiments, much less to observers. Besides, astrophysics takes it for granted that the theory holds in regions, such as the interior of stars, where it is impossible to make experiments. In identifying and disqualifying the subjectivist ingredients of quantum physics, the philosopher prepares the terrain for objectivist (realist) interpretations of it. In all three cases philosophers are justified in criticizing scientists, and in informing them that they have unwittingly adopted fragments of extravagant philosophies, such as the subjectivist idealism of Berkeley and Kant.

There have always been competent and useful philosophical teachers. But the bad teachers have been more numerous. Let us recall a few famous philosophers who presumed to correct science. Example 1: Kant sought to correct Newtonian celestial mechanics, in inventing a repulsive force that would balance the gravitational attraction, and thus explain the planetary orbits in an intuitive way. Example 2: Hegel rejected all the science that came after Kepler, including the mechanics of Newton and Euler, and the atomic chemistry of Dalton and Berzelius. Example 3: The intuitionist Henri Bergson wrote a book against the special theory of relativity, but he had the decency or prudence of withdrawing it from circulation. Example 4: Wittgenstein condemned the nascent cognitive neuroscience, claiming that it is dangerous (*sic*) to think that the mind has anything to do with the brain. Example 5: Popper held for several decades that evolutionary biology is a metaphysical theory, not a physical one; that mind and brain interact with one another; and that neoclassical microeconomics is true. Example 6: Jerry Fodor recently announced that Darwin did not understand the concept of natural selection.

1.3 Gatekeepers

One of the traditional tasks of philosophy has been defending its own frontiers, particularly from the incursions of theologians and esotericists. There have also been philosophers intent on keeping the independence of philosophy vis à vis science. Finally, a few have alerted against pseudophilosophy and pseudoscience. The first two tasks, of border patrolling, have been accomplished rather successfully: the religious philosophies were marginalized since the late Renaissance, and the scientific philosophies, though announced several times, are at best work in progress. In particular logical positivism, though initially seen as scientific, was far from deserving this characterization, as will be seen in Sect. 1.6.

The third task, that of warning against confusing wordplay with philosophy, and pseudoscience with science, will always be topical, because joining high-sounding words together is far easier than inventing worldviews, and because every scientific breakthrough seems to provoke an obscurantist reaction. Indeed, recall that the worse witch hunt in history started right after the Scientific Revolution; that the Counter-Enlightenment, particularly the philosophical component of German Romanticism, was a reaction against the scientism triumphant in revolutionary France; and that existentialism was born at the same time as quantum mechanics and the synthetic theory of evolution.

I submit that philosophers have the duty to denounce the postmodern wordplays that pass off as deep philosophy in some academic departments. I also submit that scientists have the duty to expose the pseudosciences, explaining why they are intellectual frauds, and sometimes commercial frauds as well. In Sect. 1.1 we quoted two long esoteric excerpts from Husserl, the grandfather of postmodernism. Masochists should enjoy reading his grand-children, in particular Jacques Derrida and Gilles Deleuze, who succeeded in improving on the verbal acrobatics of Hegel, Fichte, and Schelling, though not Heidegger's.

1.4 Wardens and Prisoners

Every philosophical school that rejects scientific novelties that endanger some of its principles is reactionary even it was progressive at the beginning. Hence it deserves been called a warden rather than a teacher. Let us recall a few cases of warden doctrines.

The late Aristotelians rejected Galileo's findings because they challenged some of Aristotle's statements. Berkeley mocked the infinitesimal calculus because it was imperfect like all newborns. Hume rejected Newton's mechanics because it went beyond appearances. Kant reinforced phenomenalism and proclaimed the impossibility of psychology and social science because he thought that they could be neither mathematized nor subjected to experimental control. Hegel opposed all the scientific novelties of his time because they contradicted his "objective logic". And Nietzsche rejected the whole of modernity, in particular science.

Comte, Mach, Duhem, Ostwald and other positivists condemned the atomic theory for postulating the existence of imperceptible entities. Bergson claimed that mathematics and the scientific method could not be applied outside physics. While he was Mussolini's minister of education, the neo-Hegelian Giovanni Gentile finished off the distinguished Italian school of mathematical logic. Edmund Husserl opposed his phenomenology (o egology) to all the sciences because they were realist. Martin Heidegger, his star pupil, rejected the whole of science for the same reason and also because "science does not think." The Soviet philosophers of the 1930s rejected mathematical logic because it is static, and condemned the relativistic and quantum theories for believing the subjectivist interpretations advanced by the logical positivists as well as by some vulgarizers. And the Nazis fought modern physics for being counter-intuitive and therefore contrary to the "Aryan soul".

None of these philosophical or pseudophilosophical attacks on the new sciences prevented their development, but all of them hindered it. In particular, neo-Kantianism slowed down the development of the social science, in inventing a wall between them and the natural sciences–as if there were no biosocial sciences, such as demography and epidemiology; in proclaiming the superiority of *Verstehen* ("understanding" or "interpretation") over explanation; and in adopting the methodological individualism that glosses over the global or emergent traits of the social systems, from the street-corner gang to the transnational corporation. The main rivals of this idealist school, the Marxists, remained marginal: they circumscribed themselves to commenting on the classics of Marxism and criticizing their rivals, instead of studying the societies that came after Marx and Engels.

Bertrand Russell wrote a critical history of Western philosophy, that circulated among scientists but was not well received by the philosophers because of his severe judgment of obscurantism. Here we shall limit ourselves to exhibiting a sample of the philosophies (or pseudophilosophies) that may be called *misosophies*, or *phobosophies*, because of their contempt, fear, or hatred of knowledge.

School	*Representative(s)*	*Main flaw*
Radical skepticism	Sextus, Feyerabend	Barren
Phenomenalism	Hume, Kant, positivism	Superficial
Holism	Hegel, Gestaltism	Hostile to analysis
Objective idealism	Plato, Leibniz, Hegel	Apriorist
Subjective idealism	Berkeley, Fichte, Husserl	Irrealist
Neo-Kantianism	Dilthey, Rickert, Cassirer	Anti-scientistic
Intuitionism	Bergson, Husserl, Scheler	Irrationalist
Irrationalism	Vitalism, postmodernism	Suicidal
Early Marxism	Marx, Engels, Plekhanov	Imprecise
Official Marxism	Lenin, Soviet professors	Dogmatic
Mysterianism	Religions, Jaspers, McGinn	Irrationalist
Linguistic philosophy	Wittgenstein, Ryle, Austin	Shallow
Possible worlds	Kripke, David Lewis	Escapist
Postmodernism	Heidegger, Derrida, Deleuze	Absurd
Pragmatism	Nietzsche, Peirce, Dewey	Utilitarian
Critical theory	Adorno, Horkheimer, Habermas	Antiscientific

It will perhaps be said hat neo-Kantianism, phenomenology and existentialism were fertile, since they inspired Max Weber (interpretive sociology), Alfred Schutz (phenomenological sociology), and Viktor Frankl (existentialist psychiatry) respectively. But Weber's approval of Dilthey via Rickert was purely verbal, since he was actually an objectivist; and, although he exaggerated the importance of symbols, Weber did not overlook material factors. As for phenomenological sociology, it was more homespun social psychology than sociology, since it confined itself to daily life (*Lebenswelt*), in particular conversation: it explicitly disregarded macrosocial processes, such as social conflicts, economic cycles, and wars. And existentialist psychiatry is inefficient in the best of cases, since it ignores the advances of biological psychiatry, which makes use of the findings of psycho-neuro-endocrino-immune-pharmacology. Nor has Wittgenstein's influence been constructive, since it was reduced to rejecting the scientific approach to the social, and to reducing the latter to conversation.

In short, none of the above-listed philosophical schools has contributed to the advancement of knowledge: all of them have been wardens; or worse, as in the case of radical skepticism, they have discouraged the scientific exploration of the world. This holds in particular for the contemporary version of radical skepticism, to wit, the constructivist-relativism popularized by the sociologists of science who opposed the scientific school of Robert K. Merton. Constructivism-relativism hinders the search for objective truths because it denies that these may exist: according to it, every existent is either an arbitrary invention or a social construction.

Finally, a doctrine deserves being called a *prisoner* if it advances the interests of a religious, political, or economic group. Its followers do not search for the truth because they believe to be already in its possession. They are only interested in propagating the faith, defending it from its critics, and fighting the infidels. They distrust all the ideas that circulate outside their own group. In this regard they behave like the members of a primitive tribe: they stick to tradition and banish or punish all those who adopt foreign customs.

The best-known cases of prisoner philosophies are Thomism, Marxism, and conservative political philosophy. All of them have assumed the role of keepers of the flame, and have denounced, excommunicated or persecuted the "deviationists". Thus they deprived themselves of the joy of discovery, invention, or education for innovation. In sum, they attempted to stop progress, and consequently remained behind the sciences. For example, the German philosophers contributed nothing to the sensational growth of German science between Napoleon and Hitler; but a few phobosophers, from Nietzsche to Heidegger, contributed to its equally sensational downfall in 1933.

1.5 Cheated

I shall call 'cheated' (or 'cuckold') a doctrine that feels an unrequited love for a discipline. Two clear cases, mentioned earlier, are positivism and dialectical materialism, both of which proclaim their love of science. Indeed, the sciences fit neither the phenomenalism inherent I positivism, nor the Marxist dialectics: they go far beyond appearances, and they see cooperation–without which no systems can exist–as well as the conflicts generate by the scarcity of resources, be it of a reagent in a chemical reaction, or of a piece of land in a human group.

Pragmatism too may be regarded as a cheated philosophy, this time by modern technology. Indeed, contrary to traditional craftsmanship, contemporary technology is based on science, so that it does not give priority to praxis except as the ultimate test of efficacy. This is why pragmatism is not a suitable philosophy of modern technology.

A fourth example of unfaithfulness is what may be called *panlogicism*, or *logical imperialism*. This is the belief that logic is both necessary and sufficient for philosophizing. The scholars who adopt this strategy believe that they are equipped to tackle any problem with the sole help of logic, even though this discipline is topic-neutral. In fact, some logicians have pontificated about subjects requiring knowledge they lack, from quantum physics to historiography. But whoever adopts the logical point of view will only see skeletons, because this is what logical analysis is expected to yield, namely form. For instance, he will see "a>b" beneath both "a is preferable to b" and "a is later than b". In Chap. 19 we shall discuss an example of panlogicism: the confusion between material and ideal existence.

1.6 Mercenary

Genuine philosophical investigation is just as disinterested as mathematical work: it is not self-centered and is not done primarily for profit or power, not even for the sake of any cause other than that of finding the truth. In other words, philosophizing is just as spiritual an activity as proving theorems or making music.

However, philosophical scholastics and militancy, that seek to defend or propagate a doctrine rather than an analyzing ideas or searching for new truths, are just as prosperous nowadays as in the Christian Middle Ages, when philosophy was explicitly regarded as the servant of the dominant religion.

For example, during the Cold War there were two *Hegel-Gesellschaften*, each with its own yearbook: one in the German Federal Republic and the other in the German Democratic Republic, and each dedicated to milking Hegel for its own political cause. This is understandable because, in his *Philosophy of Right*, Hegel had exalted the State, which he saw as no less than "the march of God in the world". The very same philosopher had also embraced Hobbes' legal positivism, in holding that "world history is the world court of law"–that is, might makes right. No doubt, this thesis is historically defensible; but it does not work as a philosophical foundation of law. Worse, it is a piece of court philosophy. And court philosophers deserve even less respect than court jesters, for these, unlike the former, dared speak the truth.

What more mercenary than a philosophy that, like those of Friedrich Nietzsche, Hans Kelsen, Carl Schmitt, and H. A. L. Hart, do not distinguish evil from good, and deny that there may be justice outside law? They can only compare with the "rational egoism" preached by the pop philosopher Ayn Rand and practiced by the establishment economists.

Finally, let us not forget that, in addition to intellectual mercenaries, there have been obedient philosophers, such as those who fought in the Cold War marching in step with either the CIA or the KGB. The philosopher loyal to his avocation breaks files, flees, changes job, or drinks the hemlock.

In sum, the mercenary doctrines exchange truth for servility, so that they do not deserve been called philosophies, just as courtiers are not statesmen.

1.7 Escapist

Let us call 'escapist philosophies' those that disregard he most important and interesting philosophical problems. The most obvious recent cases of philosophical escapism are phenomenology, linguistic philosophy, and modal metaphysics. The first is escapist because it is centered in the self; the second, because it denies the existence of philosophical problems; and the third for concocting pseudoproblems, such as the fate of proper names when jumping among worlds.

My complaint over linguistic philosophy is not that it is analytic but, quite the contrary, that its analyses are superficial because they make no use of any potent analytical tool, such as the predicate calculus and abstract algebra. For example, the concept of part is not analyzed by finding out how it is used in different human groups: contrary to Wittgenstein's claim, use does not make meaning but just the opposite.

Semigroup theory allows us to construct the following rigorous concept of part. Let $\oplus$ be a binary operation in a set S of objects of any kind, such as bodies or words. The concatenation $x \oplus y$ of the elements x and y is usually called 'the mereological sum' of x and y. We stipulate that, if x and y belong to S, then x is a part of y, or $x<y$ for short, if x adds nothing to y, that is, if the concatenation of x and y equals y. In symbols, $x<y=(x \oplus y=y)$.

I also suggest that analysis is but a means: that the ultimate goal of philosophizing is philosophical synthesis, that is, the construction of philosophical theories, such as theories of space time and of civic legitimacy. And I add that, since the meaning of an idea can only be found by discovering what implies it and what it implies, the best analysis is synthesis, in particular the placing of the idea to be analyzed in a hypothetic-deductive system. For example, the concept of time is not elucidated by finding out how the word 'time' is used in a given tribe, not even the tribe of physicists, but by building a theory of time or, even better, of spacetime.

As for the third of the escapist philosophies mentioned earlier, namely that of possible worlds, it cannot help explain reality since it does not deal with it. Its initial motivation was to find a use for the modal logics, that is, the formal theories of possibility. Saul Kripke, the father of possible worlds metaphysics, suggested the following interpretation of "p is possible": there is a world where p is true. No clue is provided for verification in worlds other than ours.

This interpretation has the advantage that it opens the door to unchecked imagination–that of theologians, not scientists. For example, if someone were to claim that pigs can fly, instead of correcting her, she would be told: You are right, there is a world, called Porcalia, where pigs fly. And if someone were to object that pigs cannot possibly fly because they are wingless, he would be reassured that this does not matter, since the proposition that only winged creatures can fly is not logically necessary. The theory does not place restrictions of any kind, except for consistency. It does not even define the key concept of possible world. In that theory anything, save contradiction, is possible, hence anything goes. The world is an opium (or marihuana) den.

A favorite topic of the possible worlds theorists is that of zombies, humans without a mental life that people Haitian folklore. They argue that there may be zombies because it is not logically necessary that ideation be a brain process. By the same logic, one could argue that there could be worlds with tables without legs, perpetual motion machines, gods, and the like.

In sum, apart from committing suicide, there are at least three ways in which philosophers can forget reality: bracketing it out (the phenomenological "method" or *epokhé*), focusing on words (linguistic philosophy), and fantasizing impossibles (possible worlds). Abbots and psychiatrists have always known of people who try to withdraw from the world, but never before in such numbers and not in the name of philosophy.

It may be thought that escapism is inoffensive, but this is not true. When practiced in solitude and in excess, escapism is just as self-destructive as the abuse of alcohol, television, or Internet. And when taught from the chair, escapist doctrines deviate the student's attention from interesting problems, which are both more demanding and more productive. Only parasites can afford to spend their lives day-dreaming.

1.8 Ambivalent

There are ambivalent philosophies: progressive in some regards and regressive in others. Let us briefly recall three of them: Platonism, Aristotelianism, and Marxism. Plato discarded the study of nature because he believed that the "sublunar" (terrestrial) world, being "corruptible" (changeable), is not subject to law. By contrast, he exalted mathematics, the realm of order, and proclaimed its independence with regard to the world. This is how objective and rationalist idealism, from Plato to Leibniz to Bolzano to Frege to Russell, made a huge contribution to the development and prestige of mathematics, at the same time that it hindered the advancement of the sciences of reality.

Aristotle, by contrast, encouraged all the sciences, and practiced biology and politology. Moreover, he combined his master's rationality with a timid version of the Pre-Socratic materialism. His students Theophrastus and Alexander of Aphrodisia reinforced their master's naturalism. But shortly after them Aristotelianism, although it always sought to counteract the obscurantism and subjectivism of Augustine and the neo-Platonists, became ossified: it was turned into barren scholasticism. Only the new science of Galileo, Harvey, Vesalius, Kepler, Boyle, Huygens and the members of the Accademia del Cimento–all of them rationalists, naturalists, realists, and scientistic– succeeded in overcoming ossified Aristotelianism.

Another example: Marx and Engels renewed the social studies and political philosophy while studying the industrial capitalism born only half a century earlier, which they praised for its advances while criticizing for its disequilibria and unfairness. Unfortunately they contaminated their scientific work with Hegel's holism ("historicism"), dialectics, and prophetic mania, as well as with the ideology of the "dictatorship of the proletariat."

The Marxist philosophers fared even worse: they ossified what had been a coarse but fresh and rebellious philosophy, and initially they opposed all the scientific breakthroughs of the early twentieth century, from the relativistic and quantum theories to genetics and sociology. Fortunately, there were some important exceptions: the archaeologists, anthropologists and historians who used historical (not dialectical) materialism to make some important contributions. They succeeded because they started by asking how the people they studied made a living.

Rousseau, Hume and the positivists were other important cases of ambivalence. The first two attacked the natural sciences but they held progressive views in politics and religion respectively. As for positivism, it attempted to cut the wings of physics, but in social studies it helped detach science from empty wordiness and wild speculation. The intuitionist Henri Bergson wrote reasonable pages about the roots of morals and religion; and, upon leaving Husserl's camp, Nicolai Hartmann said some interesting things about categories and levels of organization.

In short, before praising or condemning a philosophical school as a whole, let us see whether it has left something positive. After all, it may well be that in philosophy there are not gold seams but only some gold nuggets. The good goldsmith crafts diadems out of them.

1.9 Conclusion

It is well known that, even though they differ in their referents, the sciences are methodologically one: all of them seek truths and generality through clarity, rigor, and test. What a contrast between the unity and harmony of the sciences and the cacophony of the philosophies! Each philosopher tends to define our discipline in his own way, so that he may disqualify the doctrines he dislikes. But only few scholars have proposed an objective criterion to evaluate philosophies. The preceding pages suggest one: *By their fruits ye shall know them.*

Chapter 2
The Philosophical Matrix of Scientific Progress

The title of this chapter presupposes that there can be scientific progress. But half a century ago Thomas Kuhn famously attacked this assumption when he claimed that all past knowledge turned out to be false, and moreover was local rather than universal. The many constructivist-relativists who followed in his wake went much farther: they held that all the objects or referents of scientific ideas, from molecules to galaxies, are social constructions, and are therefore confined to the scientific communities that make them up (e.g., Latour and Woolgar 1979*).*

It did no occur to the constructivist-relativists that, if their thesis were true, no empirical tests would be needed: it would suffice to circulate questionnaires, asking people what they believe. Nor did it occur to them to wonder why all ambitious scientists and technologists wish to contribute to the advancement of knowledge—a progress that the relativists deny because they do not believe that there can be objective and universal truths. Indeed, they rank scientific hypotheses with tribal superstitions, and believe that they may change just as arbitrarily as clothes fashions. Let us examine this modern version of radical skepticism.

2.1 From Skepticism to Mysterianism

Instead of claiming that all the past beliefs turned out to be false, Kuhn ought to have held that, while some propositions turned out to be utterly false, others were shown to be partially true. For example, surely Pythagoras was not far of the mark when he surmised that the Earth is spherical. Later astronomers found closer approximations to the true figure of our planet, and the last word on this is unlikely to be ever uttered.

Those who do not believe that scientific progress is possible won't attempt to increase the accuracy, breadth or depth of any piece of our background knowledge, and won't even admit the existence of such fund of knowledge. Thus, whoever sticks by Kuhn's opinion will do postmodern defeatism rather than science. Nor will the Kuhnian be able to valuate projects or results, since a good scale is well designed

M. Bunge, *Evaluating Philosophies*, Boston Studies in the Philosophy of Science 295, DOI 10.1007/978-94-007-4408-0_2, © Springer Science+Business Media Dordrecht 2012

and calibrated, and comes together with a set of standard weights, whose weight is not subject to doubt.

A much older version of epistemic defeatism is the radical skepticism that Sextus Empiricus proposed in antiquity, and Imre Lakatos, David Miller and other students of Karl Popper have defended (see Stove 1982). More recently, the famous linguist and political commentator Noam Chomsky (2009) held that nature is full of mysteries and, in particular, that we shall never know what matter is, nor shall be able to explain free will. How does he know this? Because that is what some people held in the century before last. However, let us leave epistemic gloom behind, and get on with our task.

Let us start by recalling that scientific research, like any other job, can be routine or original; and that innovation, in any field, breaks with tradition in some respect while continuing it in others. It is not enough to break with the past in order to make progress: one must also attempt to replace it with something better. That is, criticism is not fruitful unless it helps conceiving of better ideas or procedures. To use rural terms, weeding is not enough: one must also dig, seed, fertilize, select, trim, and water.

Our task is to find out which conditions favor innovation in the factual or empirical sciences, whether natural, social, or biosocial. It may be argued that there are three groups of conditions of original scientific research: psychological, social, and philosophical. The main psychological conditions are talent, curiosity, motivation, intellectual courage, industry, discipline, constancy, integrity, and willingness to cooperate. The social conditions are resources, research freedom, organization, continuity and a critical mass of the research teams.

All this is rather well known, if not by politicians and bureaucrats, at least by scientists. What is far less well known is that scientific research, or at least grand research, can also be either stimulated or hindered by a philosophical outlook. For example, irrationalism and subjectivism are lethal to any rational project of exploration of reality. Such a project can only prosper in a philosophical matrix favorable to rationality and objectivity. There is more to it, and the aim of this chapter is to elaborate the above sketch. But before tackling our problem it will be well to note that the philosophical matrix is in turn embedded in a far wider matrix: the social one.

2.2 The Social Matrix

All human activities take place within a social frame. For example, business and politics require an institutional framework. Likewise, science does not prosper wherever thought is subject to ecclesiastic or political authority. The cause is obvious: Religion and the authoritarian state are conservative and therefore dogmatic, whereas science is critical and therefore innovating.

For example, the atomic theory, the most tangible and endurable fruit of ancient materialism, was suppressed by the Christian churches during more than one millennium (de la Llosa 2000); and genetics and sociology were suppressed by Stalinism (Graham 1981). Only recently, pope Benedict XVI reminded us that

Christian doctrine is inimical to evolutionary biology. And President George W. Bush restricted research in evolutionary biology and, from his high chair, recommended teaching the doctrine of "intelligent design."

However, let us not kid ourselves: nowadays the strictest censorship is the one that the scientists and scholars exert when deciding that a text is publishable. This censorship is almost always healthy in mathematics and natural science, as its aim is to prevent the spread of error and triviality. But in the social studies and humanities intellectual censorship often passes postmodern garbage while stopping originality and ideological heterodoxy. For example, in philosophy it is easier to publish scholastic comments on comments, or even existentialist nonsense, than original ideas about important but neglected problems.

Another two conditions necessary for the emergence and development of science are creative idleness, and respect for knowledge and dissidence. Already Aristotle had noted that the cultivation of curiosity requires the existence of a class of people with resources enough to enjoy idleness. But idleness is not enough, as it can be filled with chess or reading old books; nor is curiosity sufficient, because it may be satisfied without plan or persistence. In sum, the social matrix of science may be symbolized as a figure with three sides: creative idleness, freedom of thought, and resources.

Tolerance to scientific innovation is not enough: it must be stimulated and funded. We are no longer in the seventeenth or eighteenth centuries, when a well to do amateur, such as Descartes, Cavendish, or Franklin, could afford to think and experiment on his own. Doing science today may cost a lot of money in addition to talent, and it is never a solitary endeavor. For example, one hour at a particle accelerator or a great telescope may cost several million dollars. Only a wealthy state can finance such operations. And a government with a myopic science policy will let research decay for lack of funds.

It used to be said that peoples have to opt between butter and cannons. Now we must opt between missiles and ideas. And at all times since the beginning of civilization it has been necessary to opt between government for the people and against the people, between military might and social advancement. If we want science we'll have to pay for it, knowing that the only sure investments are those in research, education, and health. Those who do not invest in science will not reap technological dividends. And those who demand from scientists that they limit their inquiries to search for useful results will ask for fraud. The greatest service scientists can perform is to find truths and spread them. If, to obtain research grants, they promise that everything they will find will benefit industry or the state, they will lie—and will miss the opportunity to educate the bureaucrats.

2.3 The Role of Philosophy in the Birth of Modern Science

Why was science not born in China, the most advanced society around 1600, the date of birth of modern science in Western Europe? Joseph Needham (1956), the pioneering student of Chinese science and technology, told us why: Because the Chinese intellectual culture was dominated by three ideologies that were indifferent or even hostile to the study of nature, namely Buddhism, Confucianism, and Taoism. The Buddha, or at least his most influential disciples, had taught that all is appearance; Confucius, that what matters most is peaceful coexistence and obedience to the establishment; and Lao-tzu, that inner contemplation and religious piety trump action. None of these sages encouraged people to explore the unknown, much less to improve on the known: all three were uncurious and passive, whereas scientists and technologists are eminently curious and enterprising.

An easy way of tackling our problem is to recall how modern science emerged and, with it, five centuries of almost uninterrupted scientific progress. How and why did modern science and technology emerge only in a handful of Western European countries around 1600? We do not yet know for sure. I suppose that our ignorance of this matter is due to the fact that each historian of science has adopted a sectoral approach, looking at only one of the various progenitors of the Scientific Revolution: the Renaissance, the Reformation, the discovery and looting of the New World, the world-wide expansion of capitalism, the printing press and the accompanying spread of literacy, and the new worldly philosophy. In accordance with my systemic world-view (Bunge 1979), I assume that all of the above mentioned factors were about equally important, and that what was decisive was that they operated at the same time. However, my task in the present chapter is to focus on only one of them, the new philosophy, because there is no consensus on either what it was or why it mattered.

The Scientific Revolution is usually understood as a set of amazing discoveries and inventions that started a new and fruitful tradition. I submit that this construal amounts to beginning at the end, since nothing can be achieved without first adopting or inventing an approach, designing both a project and a strategy to carry it out, and a system of values to assess its results.

I take it that the Scientific Revolution resulted from disenchanting the world (to use Max Weber's phrase), and wanting to inventory, understand, and master it in purely naturalistic terms and with the sole help of reason and experience. In sum, the novelties generated by the Scientific Revolution constitute but a link in the following chain:

Approach – Plan – Research – Results – Evaluation.

The official story is that the new science was a byproduct of the rejection of Aristotelian scholasticism. I submit the truism that the authentic revolutions of all kinds are not restricted to removing the old: they also build new scaffoldings. In the case of scientific breakthroughs, such new scaffoldings consist in original ways of posing and tackling epistemic problems, that is, in new research strategies.

For example, Copernicus fathered the modern planetary astronomy when he replaced Ptolemy's study of individual planets with the investigation of the solar *system*. (But he kept of course two essential traits of Ptolemy's: his concern for observational data, and his use of mathematics—which reminds us that, Kuhn and Feyerabend notwithstanding, no revolution is total.) In the next century, Harvey and Vesalius inaugurated modern human anatomy and physiology when they replaced the study of separate organs with that of *systems*, such as the cardiovascular system. Newton was the first to view the universe as a *system* of bodies held together by gravitation. Quesnay and Leontieff, separated by two centuries, earned their fame by conceiving of the national economy as a *system*, not just a set of individuals. Faraday and Maxwell thought of electric charges and currents, together with magnets, as *systems* held together by electric and magnetic fields.

Darwin conceived of biospecies not as separate collections of organisms but as branches of the tree of life—a *system* whose components are bound by the descent relation. Ramón y Cajal revolutionized neuroscience when he identified the units of the nervous tissue and revealed the networks or *systems* into which they combine. Rutherford and Bohr fathered modern atomic physics when they conceived of the atom as a *system* of particles rather than a tiny marble. Bernardo Houssay was awarded the Nobel Prize for proving that the pancreas and the hypophysis or pituitary, though rather distant from one another, belong to the endocrine *system*. Donald Hebb, inspired by the work of Cajal and his school, renewed psychology upon postulating that what feels, perceives, thinks or evaluates are neither whole brains nor individual neurons but something in between: cell assemblies or neuronal *systems*. And John Maynard Keynes created modern macroeconomics, as well as the socioeconomic policy that bears his name, by conceiving of the economy of a region as a social *system* closely connected to the political system, rather than as an amorphous aggregate of free agents—the way the neoclassicals treated it.

In all of the above-mentioned cases the innovation was centered on a *material system* rather than on either an isolated individual or an impenetrable whole: Planets were no longer thought to be pushed by angels, neuronal systems were not deemed to be controlled by immaterial souls, economies were not under the guidance of invisible hands, and so on. In sum, the makers of modern science were tacitly practicing what I call *systemic materialism*, an alternative to both physicalism and dialectical materialism. The central thesis of this ontology is that very real entity, whether atom or cell, animal or society, is a material system or a component of such. This kind of materialism does not eliminate the mind: it only regards mental processes as brain processes, and consequently as subject to normal scientific investigation. Nor does systemic materialism ignore or underrate the symbolic traits of cultures: it just conceives of them as products of humans living together, on a par with labor and conflict. However, let us get on with our story.

I also suggest that the Scientific Revolution consisted to a large extent in developing three capital aspects of Aristotelianism: rationalism, realism, and the cautious materialism of Theophrastus, Alexander of Aphrodisias, Averroes, and the Latin Averroists. Indeed, the schoolmen respected logic although, as Bacon and other critics rightly noted, they used it to dispute over the sacred scriptures rather than to explore the world. And, except for the nominalists—who were actually vulgar materialists—the schoolmen replaced naturalism with a moderate supernaturalism that did not interfere seriously with physics. For instance, Tomas Aquinas, in defiance of the establishment, promoted the study of Aristotle's physics, which had no trace of supernaturalism.

Moreover, the Medieval theologians had not challenged the epistemological realism inherited from Greek antiquity. The challenge came only when Copernicus and his followers proposed the heliocentric model of the planetary system. Since this model contradicted the book of Genesis, Cardinal Bellarmino and other Catholic theologians came up with an ingenious trick: conventionalism (or instrumentalism). According to this epistemology, scientific theories are neither true nor false, but only descriptions of appearances and tools for making predictions. This was said to hold, in particular, for the heliocentric and geocentric models of the "world" (the planetary system). In fact, both were (at that time) compatible with the extant astronomical data, so that a good empiricist was not compelled to choose between them. In short, they were empirically equivalent models. Ironically, three centuries later Philip Frank and other logical positivists resurrected this doctrine and thus tacitly approved of the pivotal accusation of the Inquisition against the realist Galileo.

This episode illustrates the difference between realism and empiricism: Whereas realists attempt to account for reality, empiricists—from Ptolemy to Hume, Kant, Comte, Mach, and Carnap—stick to phenomena or appearances. Thus they are anthropocentric, since there can be no appearances without subjects. The fact that most empiricists have declared their love of science is beside the point: the point is that empiricism is an obstacle to scientific research because it is subject-centered, whereas scientists attempt to be as objective as possible.

In summary, the progenitors of the Scientific Revolution practiced the critical rationalism, naturalism (or even materialism), and epistemological realism inherited from Greek antiquity, as well as the systemism that emerged along with modernity. Let us take a closer look at these philosophical ingredients of modern science.

2.4 Materialism, Systemism, Dynamicism, and Realism

The Venetians knew that religious and philosophical tolerance is good for business, particularly for foreign trade, the source of their great wealth. For this reason they fully respected academic freedom at the Padua University, which they funded and sheltered from the Inquisition. About 1600, its most famous professor was Cesare Cremonini (1550–1631), a colleague, friend, and rival of Galileo's. Cremonini was the most famous and best-paid philosopher of his time, as well as a correspondent and protégé of several European princes. He was also a notorious heretic, whom Torquemada would have liked to burn at the stake because he was an outspoken materialist, realist, and rationalist, as well as a faithful follower of Aristotle. Cremonini was popular among the Paduan students for denying the immortality of the soul and holding that reason trumps faith (see Renen 1949).

What was the famous materialist and rationalist professors's contribution to the Scientific Revolution headed by his colleague Galileo and his former student William Harvey? None, except for the unfortunate role that Galileo assigned him in his soon to be condemned *Dialogue*. In fact, Cremonini served him as a model for Simplicius, one of the two schoolmen who refused to look at the Sun and the Moon

Fig. 2.1 Increasing specialization (analysis) is balanced by ever broader integrations (syntheses)

through the telescope. Why bother looking at them, if the Philosopher had described them as perfect spheres, whereas Galileo claimed that the Sun was spotted, and the Moon was blemished by craters? (Today's analog: the philosophers of mind who refuse to look at cognitive neuroscience.)

A moral of this episode is that, to engage in scientific research other than routine work, materialism, realism and rationalism are insufficient. That is, it is not enough to deny the existence of disembodied spirits, or to affirm the reality and knowability of the external world. To do science it is also necessary to adopt the scientific method, that is, to check one's conjectures against the relevant data as well as with the theories enjoying a good track record. Shorter: scientism is a necessary condition for doing good science.

Nor is scientism enough to do grand science: a systemic approach too is required. That is, one must try and place the object of inquiry in its context instead of isolating it from its environment. For example, Walter Cannon, Hans Selye and their successors are remembered for having built the interscience that studies the supersystem constituted by the nervous system and the endocrine and immune systems, namely psycho-neuro-endocrino-immunology. Other young interdisciplines are cognitive neuroscience and evolutionary developmental biology, or evo-devo for short. Similar fusions have emerged in the social sciences in recent years, such as socioeconomics and political sociology. (Sociobiology and its successor, evolutionary psychology, do not count because they are unsuccessful attempts to reduce social science to biology.) All of the above fusions, and then some, were effected one century after the birth of physical chemistry (or chemical physics), biochemistry, biophysics, and social medicine (née social hygiene).

In sum, it has become increasingly clear that the disciplinary divisions are largely artificial, because the universe is the system of all systems. The methodological consequence is that the specialization or disciplinary division process should be balanced by a parallel process of progressive fusions: from the trunk to the branches and back (Bunge 2003). See Fig. 2.1.

Systemism somehow entails dynamicism (or process metaphysics), because every interaction causes changes, both internal and external. Dynamicism is contrary to the Platonic worldview, according to which the basic constituents of the world are immutable, disembodied, and eternal ideas. Ironically, Plato, an objective idealist, gave the correct definition of 'material' as 'changeable' (or 'corruptible', in his hierarchical worldview). Any materialist who admits this identity of materiality with changeability will be a dynamicist. That is, she will hold that whatever exists really is impermanent: it is changing or will eventually change. In particular, biologists will think in developmental and evolutionary terms, and anthropologists in terms of biosocial evolution.

A dynamicist ontology will favor the search for the laws of change as well as for the mechanisms of complex things, that is, their characteristic processes, such as reproduction, work, and trade in the case of human groups. It is no accident that Machiavelli, the founder of modern political science, explicitly criticized Plato's thesis that whatever changes is imperfect. Machiavelli saw society in constant change. Nor is it accidental that Galileo sought the laws of motion both experimentally and mathematically instead of repeating beliefs of his predecessors.

Since then, the kernel of any advanced scientific theory is a set of equations of change. Since then, too, nearly all scientists expect that today's best theories will eventually be perfected or replaced: only unreconstructed Marxists, orthodox economists, and arrogant physicists believe that their own theories, or at worst the next ones, are or will be final. Scientism involves both moderate skepticism about current knowledge and meliorism, or trust that future research is bound to produce new and more accurate or deeper knowledge.

2.5 First Parenthesis: The Ossification of Philosophy

Let us now jump three centuries, from Rome in 1630, to Moscow in 1930. At that time, the USSR had tens of thousands of little Cremoninis, namely, the professors of Marxist philosophy who called themselves materialists, as well as friends of reason and science. They also passed off as realists because they did not distinguish between realism, an epistemological school, and materialism, an ontological one. (The root of this confusion may be Lenin's definition of 'material' as whatever exists outside consciousness—surely a sufficient but not necessary condition.)

Those professors of Marxist philosophy were not original thinkers but catechists, commentators, and keepers of the flame. Worse, they attacked systematically all the great scientific novelties of their own time, in particular relativistic and quantum physics, genetics, the synthetic theory of evolution, and even mathematical logic. Those new Cremoninis, none of whom produced any scientific or philosophical novelty, held the scientific theories they condemned were false, because they believed that they contradicted the official philosophy.

It seems that the Soviet Marxists opposed mathematical logic because they shared Hegel's conflation between logic and ontology. They believed that logic, like factual science, had to "mirror" the world instead of analyzing ideas. And they opposed relativity, the quantum theory, and the new biology because they had no first-hand access to them, since actually all three were perfectly materialist and realist. However, this aspect of the question is hardly relevant to our problem. The point is that the Marxist schoolmen were not scientistic: They recited canonical texts and praised science from afar instead of looking closely at what scientists did.

Half a century later I got involved in a similar controversy, though on a much smaller scale. In the World Congress of Philosophy held in Düsseldorf in 1978

I held a public debate with Sir John Eccles, the famous neuroscientist, a Nobel laureate, and collaborator of our common friend, the philosopher Sir Karl Popper. Eccles, who has been called 'the ecclesiastical neuroscientist', held the idealist (and parapsychological) view that the immaterial mind keeps the neurons moving, much as the pianist plays the piano. By contrast, I held just the opposite: that minding is the specific function of the human brain. Eccles did not invoke any experiments, whereas I supported my thesis on the then nascent experimental cognitive neuroscience. The local newspaper chronicled the event on the front page. Eccles warned me that he would retaliate the next day, after my own paper with Rodolfo Llinas. But, although he sat on the first row, he did not open his mouth.

The story did not end there: Eccles got an ally from the place he least expected and wished. In fact, during the same congress someone introduced me to the editor of the Russian journal *Filosofskie Naukie*, who asked me for a paper. Evenrually I mailed him my paper "The bankruptcy of psychoneural dualism". The journal published it in 1979, followed by a criticism, twice as long, by D. I. Dubrovsky. To refute me, this individual invoked the authority of Lenin, who, in his very interesting *Materialism and Empirio-criticism* (1908), had criticized Joseph Dietzgen, the tanner and self-taught philosopher whom Marx had called "our philosopher". Dietzgen had held that ideas are just as material as motion and digestion, because they are brain processes.

Lenin argued that, if thought were material, then there would be no conflict between materialism and idealism. He did not realize that, by denying the materiality of the mental, he unwittingly embraced psychoneural dualism. The same point escaped his sycophants in the so-called "real socialism", who defended instead Pavlov's "second system of signals", alien to experimental psychology. Thus it came to pass that cognitive neuroscience, which is inspired in psychoneural monism, was born outside the Soviet empire. This opportunity was then missed by a combination of dogmatism and conceptual confusion with ignorance of science. In turn, this combination might not have occurred if dialectical materialism had included scientism instead of Hegel's hermetic dialectics—in which case it might have deserved being called 'scientific materialism'.

How can the ossification of a philosophy be avoided? By working on problems rather than authors, and by combining rational debate with research and mixing with other tribes—those of mathematicians, scientists, and technologists. In turn, such mixing is favored by organizing multidisciplinary seminars attended by researchers in all fields. And this for three reasons: because scientists are the ones who excel in the use of the scientific method—even if they profess not to believe in its existence; because the interdisciplinary frontiers are partly arbitrary; and because there is no justification for philosophizing in total ignorance of what is going on outside philosophy. Indeed, all of the so-called perennial problems of philosophy, such as what are the mathematical objects, matter, space, time, causation, chance, life, mind, society, and history, are worked out in detail by researchers in the various special sciences. However, let us get on with our task of sketching the philosophical matrix of science.

2.6 Scientism, Rationalism, and Humanism

To innovate in the young sciences it is necessary to adopt scientism. This is the methodological thesis that the best way of exploring reality is to adopt the scientific method, which may be boiled down to the rule "Check your guesses." Scientism has been explicitly opposed by dogmatists and obscurantists of all stripes, such as the neoliberal ideologist Friedrich von Hayek and the "critical theorist" Jürgen Habermas, a ponderous writer who managed to amalgamate Hegel, Marx, and Freud, and decreed that "science is the ideology of late capitalism."

By contrast, Lalande's sober *Vocabulaire* (1938: II, 740) gave the following definition of *scientism*: "it is the idea that the spirit and the methods of science ought to be extended to all the domains of intellectual and moral [social] life without exception." Thus, contrary to its detractors, scientism is not the same as social naturalism, or the attempt to ape the natural sciences in the social domain: it is only the attempt to use the scientific method in dealing with all the problems concerning facts. This is, for example, the approach of the scientist who, upon finding an unusual artifact in an archaeological site, makes conjectures about its possible uses, and subjects them to tests: he engages in experimental archaeology.

Scientism was not born until the end of the Enlightenment. Ancient India produced as many philosophical schools as Europe (see Dragonetti and Tola 2004). In particular, it gave materialist, realist, and humanist philosophers. But until a century ago there was no scientism in India because there was no science either. Science generates scientism, which in turn promotes science.

Scientism is opposed to irrationalism, in particular intuitionism and mysterism—the claim that there are mysteries, such as the nature of the mind, that science will never solve. But scientism is also rejected by those who try to pass off their improvised speculations as scientific findings. This is the case with the selfish gene idea, memetics, psychological nativism, and speculative evolutionary psychology. The latter is the attempt to explain everything social in exclusively biological terms, and moreover by implausible hypotheses, such as that we are walking fossils: that our minds have not changed over the last 50,000 years. Moreover, evolutionary psychologists overlook the fact that there are social inventions, and that some of these—such as warfare, human sacrifice, some dietary rules and some kinship rules in pristine societies—are detrimental to life.

Scientism includes rationality, that is, the requirements of clarity and logical consistency. The least we can expect from a philosopher is clear expression and reasons for or against the theses she discusses. A hermetic doctrine, one so meaningless that it is not even false, so that it cannot be rationally debated, does not deserve to be called a philosophy.

It may be objected that rationality is desideratum rather than fact; that psychoanalysis, behaviorism, the computer model of the mind, and evolutionary psychology, have shown that we are basically irrational: that we are slaves of passions (Hume, Nietzsche, Freud) of innate programs, or living fossils, whose minds were shaped during the Pleistocene to cope with the dangers of the African savannah. These doctrines, that John Kihlstrom called "stupidist", have been severely criticized by Elliot Turiel 2010). This researcher reminds us that, from Jean Piaget on, developmental psychology has piled up evidence for Aristotle's hypothesis that we are rational (or at least reasoning) animals, although it is true that reasons intertwine with emotions and actions. In sum, the rationality postulate is descriptive as well as normative.

Scientism is centered in rationality but it excludes dogmatic rationalism or apriorism, according to which logic suffices to understand reality. The reason is that logic is topic-neutral: it makes no assumptions about real existents. Scientism also excludes the paraconsistent logics. These are formal theories that admit contradiction and therefore reject the kernel of rationality, so that they place themselves beyond he pale of logic proper. Indeed, from the start the very raison d'être of logic has been the protection of consistency, a necessary condition for cogent deliberation and rational action.

Newton da Costa's original motivation for proposing paraconsistent logic is Hegel's thesis that the world is "contradictory", that is, ridden with conflict. If this is true, and if logic is mistaken for ontology, it follows that logic must admit contradiction proper, that is, logical inconsistency, which in its simplest form consists in the simultaneous affirmation and negation of a statement. But this amounts to the death of reason, without which it is impossible to manage conflict.

Besides, if we tolerate contradiction we cannot use the reductio ad absurdum argument, and thus we lose the great majority of mathematical proofs. Worse, we lose the very notion of absurdity, and by the same token we consecrate all the postmodern absurdidties. In brief, the paraconsistent logics are illogical, much as selfish morality is immoral. The least a philosopher can do is to alert us against absurdity. However let us return to scientism.

Obviously, scientism is necessary but insufficient for scientific progress. Indeed, the scientific method may be used to handle uninteresting or even ridiculous problems, that is, problems with trivial solutions, or that ignore well known pieces of knowledge. This is the case with the publications in scientific journals that have earned the Ig Nobel Prize, which is awarded every year to ten papers published in reputable science journals. The following are among the findings that deserved the 2008 prizes: Dog fleas jump higher than cat fleas; armadillos may mess up archaeological sites; the more expensive placebos are the most effective; hair tends to get tangled; and plants have dignity. As Molière said, a learned fool is even more foolish than an ignorant one.

Scientism can be faked just like science, art, love, and money. For example, during two centuries positivism called itself scientific, and was even taken for scientism. Indeed, it proclaimed its love of science at the same time that it attempted to force into the phenomenalist straitjacket, which bans imperceptible entities and properties, such as atoms, genes, mind, and social forces.

The confusion between scientism and positivism was so common during the second half of the nineteenth century, that it spilled over into politics. Indeed, between 1880 and 1910 positivism enjoyed such a prestige, that the Argentine intellectuals, all of them progressive at the time, swore by Comte and Spencer.

But these were the same heroes of the "scientific" party led by the dictator Porfirio Díaz, that paralyzed Mexican society during its long rule. Unsurprisingly, after that period both the Mexican progressives and the Argentine conservatives replaced positivism, misnamed scientism, with idealism. However, let us return to the philosophical pentagon.

To conclude the discussion of scientism, note that it is not a substitute for philosophy but a philosophical program. Philosophy is unavoidable because it deals with extremely general concepts that are used carelessly in all fields, such as those of reality, becoming, knowing, and the good. Yet, since most philosophies have dealt unsatisfactorily with those ideas, there is a tendency among scientists and technologists to believe that they can dispense with philosophy. Experience suggests that ignoring philosophy leads to using ignorant philosophy. Scientism does not deliver good philosophy but helps avoiding bad philosophy, as well as thinking philosophical theories using what is known and helping explore the unknown.

Last, but not least, scientific progress also requires observing the moral norms that control the search for truth and its dissemination. This is because scientific research is a social endeavor, as it involves cooperating in some respects while competing in others. In fact, even the most recluse of investigators utilizes findings of others, and in turn feeds his readers. And, to avoid purely destructive conflicts, all social transactions must observe norms of coexistence, which are moral norms. For example, plagiarism is more strongly condemned in the sciences than in the humanities, because teamwork is more frequent in the former than in the latter.

In a classical article, the founder of the scientific sociology of science (Merton 1975: 259) showed that basic research is ruled by the following moral norms: intellectual honesty, integrity, organized skepticism, disinterest, and impersonality. I would add a few more: cooperating with peers and students rather than exploiting them; combining research with mentoring and teaching; promoting the free and fair competition for grants, students, and jobs; not skirting problems whose research may annoy the powers that be; telling the truth even if, indeed particularly if, it contradicts he ruling worldview; popularizing science and scientism; denouncing pseudoscience and obscurantism; and abstaining from using science to harm people.

The set, or rather system, of norms ruling the search for truth and justice is often called *humanism*. Humanism amounts to a combination of social responsibility with universalism: it enjoins us to avoid whatever may harm most people while benefiting particular social groups at the expense of the majority.

Humanism condemns military aggression and racism, political oppression and cultural exclusion. The doctrines that promote aggressive nationalism, regionalism, racism, or sexism—whether masculinist or feminist—are ideological, not philosophical. This holds in particular for the diatribes of Nietzsche and Heidegger against humanism.

The exclusivist and relativist doctrines are not humanist, because they are not universalist. Moreover, they are conservative or reactionary, since they defend vested interests. Humanists proclaim the universality of the basic philosophical values—clarity, consistency, and truth—as well as of the social values of the French

Revolution of 1789: liberty, equality, fraternity. (Notice that these values consttitute a package or system: every one of them can be realizad only along with the other two.) Humanists are universalist on morals, science, and philosophy. However, they defend the right to differing in everything else, in particular occupation, lifestyle, food, clothing, technology, and politics.

2.7 Second Parenthesis: Logical Imperialism

During the first half of the last century mathematical logic expanded to such an extent that a number of logician and philosophers came to believe that it suffices for doing ontology, epistemology, and ethics, and even to rationally reconstruct scientific theories. This program may be called *logical imperialism.*

Some of its most outstanding practitioners were Stanislaw Lesniewski, Rudolf Carnap, Nelson Goodman, Patrick Suppes, Saul Kripke, Wolfgang Stegmüller, Jaako Hintikka, and David Lewis. All of them worshipped logic but ignored the sciences of reality.

The best known of the doctrines of this kind is modal metaphysics, or the metaphysics of possible worlds. Two of its leaders are Saul Kripke and David Lewis. Both believed that modal logic, which is centered in the general concept of possibility, is the key to discuss about existence, causation, mind, and other key metaphysical concepts. They do not realize that all logics are ontologically neutral. In particular, none of the 256 (possible) systems of modal logic distinguishes between conceptual and real existence and possibility. To them, "There are zombies" and "There are stones" belong in the same league; likewise "It is possible to think without a brain" and "Normal brains may learn new tricks".

In particular, Kripke does not care that psychology denies the possibility of the existence of zombies—humans like us but mindless—since, to behave as a human, it is necessary to have a brain in good state. Like the proverbial philosophers who refused to look through Galileo's telescope, Kripke maintains that here are zombies in some world just because they are conceivable; moreover, such possibility would refute the psychoneural identity hypothesis. Since his premise "There are zombies" is factually false, his arguments are not sound. Moreover, they are ridiculous, on top of which they are motivated by religion, not science. Much the same applies to Hilary Putnam's "Twin Earth," a planet that would be the same as ours, even in that it would be inhabited by humans, but would have no water—without which life is impossible. There is a name for such fantasies: parlor games.

Similar objections may be raised against the rational-choice theorists, who claim to explain everything social by a single postulate, that of economic rationality. Indeed, this postulate states that all human actions result from the attempt to maximize the expected utilities. Those theorists never wondered whether real humans behave this way. (Psychologists and behavioral economists, such as Daniel Kahneman and the Zürich school, found that most people do not behave like that.) Nor have they noticed that all the social agents belong to a number of social systems or networks, such as families, business firms or nations, characterized by global properties and constraints, that is, traits that do not belong to individuals but protect or restrict personal freedom.

Some rational choice theorists such as Gary Becker, call themselves ‘economic imperialists’, and they incur the simplistic fallacy of assuming that their referents are characterized by just a couple of traits, suh as maximizing behavior in the case of individuals, and composition in that of collectivities. This simplistic strategy has prevented them from explaining the most salient social facts, such as the emergence and dismantling of systems, cooperation and conflict, economic crises and wars, political advances and retrocessions, or the ingenious or beautiful but useless inventions, such as music and pure mathematics.

In sum, formalism, in particular logical and economic imperialism, has the advantage that it is clear and inexpensive, but the disadvantage that it trivializes and therefore discredits philosophy and science. In these fields, formal tools are a means, not an end: a means to clarify ideas, increase rigor, and facilitate the construction of theories about important matters.

2.8 The Philosophical Pentagon

The upshot of the preceding is that the search for new and important truths about reality involves a more or less explicit commitment to the philosophical pentagon made up of scientism, realism, materialism, systemism, and humanism: see Fig. 2.2.

I submit that Galileo and his disciples, as well as Kepler, Harvey, Huygens, Boyle and other founders of modern science adopted tacitly the philosophical pentagon. Descartes only missed scientism, since he did not bother to look for empirical evidence for his fantasies about the ubiquitous whirls. Much closer to us we find at least five great scientists who adopted the pentagon and changed the dominant worldview: Darwin, Cajal, Rutherford, Einstein, and Keynes—four natural scientists and one social scientist. (However, it may be argued that Marx came close, and would have qualified, had it not been for is residual Hegelianism and his fondness for prophecy.)

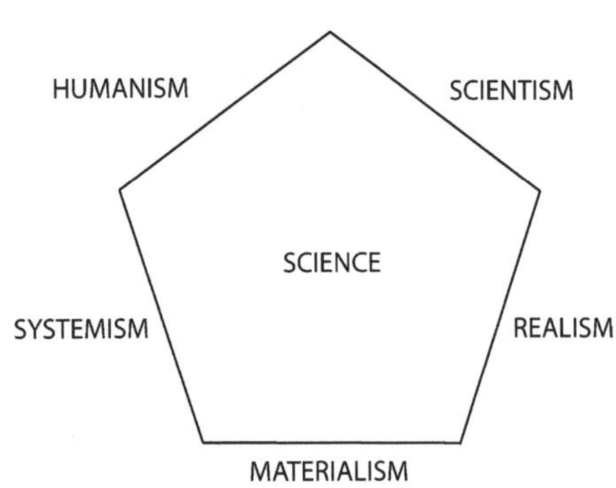

Fig. 2.2 The philosophical matrix of scientific progress (From Bunge 2010: 242)

Why that asymmetry between social and natural scientists? It cannot be a matter of age, because historiography and politology are just as old as the natural sciences. Could it be because telling truths about social issues takes more objectivity, independence and civil courage than telling them about nature? Or may it also be to the fact that many students of society have been strongly influenced by philosophers who, like Kant, Hegel, Dilthey, Bergson, Husserl, Habermas, and Putnam have denied the possibility of understanding society through the scientific method? However, this is a subject for another investigation (see, e.g., Albert 1994; Bunge 1996).

2.9 Irregular Pentagons

In most cases, the sides of our philosophical pentagon are not equal: sometimes materialism is short, others realism is insufficient, and in still others humanism is weak. For example, until a century ago many biologists were vitalists; many students of society succumbed to Kantian subjectivism; and most nineteenth century physicists opposed atomism while chemists cultivated it. Even today, one may find psychologists and social students who reject the scientific method, and maintain, against all evidence, that we are born knowing the basic rules of grammar, that criminality is inborn, that languages do not evolve, or that all markets are in equilibrium.

Paradoxically, nearly all of the founders of microphysics, in particular Bohr, Heisenberg and Born, preached positivism while investigating entities that the fathers of positivism, Comte and Mach, had declared nonexistent. In fact those giants of physics denied the independent (objective) existence of the very objects that they studied so successfully, namely atoms and their components: they extolled subjectivism and phenomenalism. But of course they did not practice the philosophy that they extolled: they did not include the observer in their equations, and they admitted the reality checks performed in the physics labs (Bunge 2010). In his case, a bad philosophy did not frustrate the ancient atomic project, of explaining the visible by the invisible. Likewise, Newton praised inductivism in the same book, his monumental *Principia*, where he expounded the earliest fruit of the hypothetic-deductive method in natural science. The great scientist does not surrender to bad philosophy. But the bad philosophy he indulges when writing for the great public may frustrate scientific callings.

Paradoxically, psychology fared in the opposite way: the psychoneural identity hypothesis, which medical psychology has been maintaining since Hippocrates and Galen, and neuroscience since Broca and Wernicke, did not lead physicians to making experiments nor to imagine models of brain functions. This is due to the fact that physicians, save exceptions, have not practiced the scientific method. Instead, they adopted the empiricist case study introduced by Hippocrates. For example, Juan Huarte de San Juan's *Examen de ingenios para las ciencias* (1575), a bestseller in several languages during a whole century, contains practically nothing that Galen did not know: Huarte repeated and popularized instead of conducting scientific research.

2.10 From Social Science to Sociotechnology

Similar mismatches between science and philosophy have happened in the social studies. For example, Marx conducted original and important scientific research while praising Hegel's antiscientific philosophy. And, as Trigger (2006) showed, a number of Marxists made original contributions to archaeology thanks to their materialist, systemic and scientistic approach. Much the same has been said of some French and British Marxist historians (Barraclough 1979). By contrast, the Marxist philosophers only complicated and codified the absurdities of dialectical ontology.

Two giants of the last century, John Maynard Keynes and Gunnar Myrdal, did original social science and sociotechnology, at a high level and with a humanist orientation. Keynes founded modern macroeconomics, and Myrdal fathered socio-economics, and both scholars inspired social policies aiming at improving the lot of the many. By contrast, their main critics, Milton Friedman and Friedrich Hayek, were orthodox—hence unrealistic—in theoretical matters, and they exhibited social insensitivity. Their bad science and regressive political philosophy may be traced back to their ontological and methodological individualism together with their uncritical epistemology.

The economic crisis that the so-called West has been suffering since the end of 2008 has shown the flaws of standard economic theory and its individualist under-pinnings, together with the moral insensitivity of major business leaders and their political helpers (see Bunge 2009). One of the falsities of that theory is the thesis that the market is naturally free, competitive and self-regulated, while everyone knows that it is distorted by monopolies and monopsonies, as well as by barriers to entry and the corrupting action of lobbyists. These and other factors result in fluctuations that only state regulations may dampen.

Interestingly, the soundest and most competitive economies are also tightly regulated and accompanied by advanced welfare states and the lowest income inequality. For example, the Gini index of Denmark and Sweden is 0.23, half that of the US and the UK (OECD 2008).

2.11 Dogmatic and Programmatic *Isms*

Every one of the sides of the philosophical pentagon (Fig. 2.2) may be taken in either of two ways: as a programmatic or as a dogmatic *ism*. Let me explain. A *programmatic ism* is a broad research program motivated by a philosophical hypoth-esis concerning the nature of the objects of interest as well as the right way to tackle them. By contrast, a *dogmatic ism* is a wall of a cage that restricts the flight of the scientific imagination. Caution: What was initially a fruitful programmatic ism will turn into a dogmatic ism if its practitioners refuse to admit that the program has hit a barrier. (Imre Lakatos made much of this, which he called the degeneration of the research program.)

Let us glance at a handful of famous programmatic *isms*.

1. *Mathematization*. Galileo's program, of mathematizing whatever can be mathematized, may be called 'mathematicism'. This is a trait of modern thought, by contrast to the premodern one, that was typically imprecise and qualitative. The success of mathematicism in physics, chemisty and engineering is beyond doubt. In other fields is has had a limited success. Worse, in some cases it has served to hide imprecision or dogma under symbols of mathematical appearance as is the case with much of mathematical economics, which has only a tenuous relation with reality.
2. *Mechanism*. The success of modern theoretical mechanics renewed the mechanistic worldview born in ancient Greece. The most advanced thinkers from Galileo to mid-nineteenth century believed that every existent was a body or a collection of particles satisfying the laws of mechanics. The mechanist project was enormously successful till the birth of field theory. In biology it succeeded with the birth of biomechanics in Galileo's own brain. Nowadays biophysics is still alive buit it cooperates with biochemistry, and the attempt to account for mind and society in mechanical terms was forgotten long ago.
3. *Informationism*. The sensacional technological and commercial achievements of information technology since mid-twentieth century suggested looking at everything in terms of information. In particular, it was claimed that information fiuxes are the essence of life, mind, and society, regardless of material things. The information enthusiasts overlooked the facts that information does not exist by itself, that the natural precedes and supports the artificial, and that information overload can be just as harmful as information scarcity.
4. *Geneticism*. The discovery o the key role of genes in elementary biochemical processes, such as protein synthesis, gave rise to the legend that "genome is destiny", and consequently that genetics is the mother of all biological sciences. In the year 2000 the Human Genome Project was hailed as the endeavour to decipher the Book of Life, and it was financed because it promised to revolutionize medicine. The project was carried out in the record time of half a year because it used well-known ideas and techniques. But it turned out to be a sink rather than the promised fountain. Indeed, a decade later the managers of the two research teams that worked on it, Craig Venter and Francis Colllins, admitted that very few of the inicial promises were kept.

   (However, the Human Genome Project had at least two unexpected consequences. One of them was that it changed many lives, for it attracted to science an uncounted number of young people, particularly girls, and it led many biologists to modifying their research projects. The second unanticipated consequence was that the cuts torced on the original ambition reinforced the idea that organisms exist at the same time on several levels—molecular, cellular, organismic, etc.—every one of which is characterized by properties absent from other levels. This results in that none of the life scieces may dominate the others, even though some of them may temporarily enjoy a stronger support due to a false (reductionist) philosophy of biology. This science is a system, so that, when any of its components weakens, all the others suffer too. In short, geneticism, a variety of radical reductionism, had its day of glory, but it is now understood that all things have to be studied at all their levels.

5. *Economism.* The central hypothesis of economism is that everything social is economic or has exclusively economic causes. (The classical formula reads "Everything in society is economic or reducible, in the last instance, to economic factors"—a characteristically irrefutable hypothesis, given that "the last instance" may be conveniently pushed forward as far as needed.) The principal merit of this *ism* is having weakened spiritualim, according to which everything is ideal or a product of ideas. But, since all social processes have environmental, political and cultural strands entwined with economic ones, economism has severey distorted all of the social sciences, economics included. For example, every economic breakhrough has had political and cultural (in particular technological) causes and effects, as well as environmental impacts.

Every one of the *isms* we have just glanced at started by being progressive as an ambitious research project, but ended up by hindering progress because it was one-sided. In general, fashions mobilize at first but end up by immobilizing—which is why they deserve to pass.

2.12 Concluding Remarks

In conclusion, the philosophical "procreant cradle" of scientific research—as Shakespeare might call it—may be either a nest to lay and incubate promising ideas, or a cage to prevent the flight of imagination. In other words, a philosophy may be progressive or reactionary, according as it facilitates or obstructs the advancement of knowledge. The natural scientists have tacitly practiced an inchoate progressive philosophy for half a millennium. This is why they have not been affected by the postmodernist fashion, which is nothing but a rehash of the Counter Enlightenment born about 1800. By contrast, the students of society and humanities are not immunized against that plague.

What can be expected from a student of society who takes seriously the spiritualist, subjectivist, and individualist philosophy that Max Weber and Pitirim Sorokin preached but hardly practiced, whereas the phenomenological sociologist Alfred Schütz and the Diltheyan anthropologist Clifford Geertz both preached and practiced? For one thing, such a student would not bother to find out how people make a living and, a fortiori, she would not care for social statistics. This is because, being a spiritualist, she would be interested exclusively in beliefs, games, and ceremonies. For the same reason, she would be unable to understand social conflicts and wars, since nearly all of these are motivated by conflicting material interests. Nor would she understand the current economic crisis, since one of its causes is the fight of big corporations and their political and academic friends against all the regulations aiming at restricting their power over individuals. And, being anti-humanist, the student in question would refuse to acknowledge that the most highly developed and

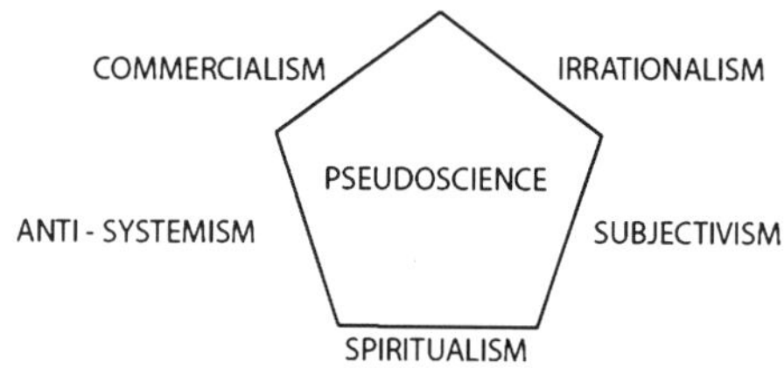

Fig. 2.3 The philosophical incubator of pseudoscience

economically successful nations are those where the economy is regulated and the welfare state flourishes. Lacking in true social theories and a humanist morality, our victim of pseudoscience would favor mean socioeconomic policies that, in the end, would worsen the crisis of old-fashioned capitalism, which is undergoing a global crisis and threatening the open society, as George Soros (1998) wrote a decade ago.

In summary, all scientific research is conducted inside a social framework and a philosophical matrix. If the social framework is broad, flexible, and encourages innovation, it will support the individuals who are motivated by curiosity and peer recognition. And if the underlying philosophy is scientistic, materialist, systemic, realist, and humanist, then disinterested research may bear fruit. By contrast, bad philosophy is likely to breed pseudoscience: see Fig. 2.3.

True, neither the principles in question nor their duals occur explicitly in any scientific argument: they are regulative, not constitutive, to employ Kant's correct distinction. Hans Reichenbach might have placed those regulative principles in the "context of discovery," not in the "context of justification." But this poses no problems except to the philosophers, from Bacon to Husserl, who demanded a context-free science, one without presuppositions. But there is no such thing, because the empty set cannot breed. At any rate, only unexamined presuppositions can harm. Which is why philosophical diggers can be useful.

References

Albert, Hans. 1994. *Kritik der reinen Hermeneutik*. Tübingen: J. C. B. Mohr (Paul Siebeck).

Barraclough, Geoffrey. 1979. *Main trends in history*, New York, London: Holmes & Meier.

Barrow, John D., Paul C.W. Davies, and Charles L. Harper Jr. (eds.). 2004. *Science and ultimate reality*. Cambridge: Cambridge University Press.

Bunge, Mario. 1979. *Treatise on basic philosophy*, *vol. 4, A world of systems*. Dordrecht: D. Reidel.

Bunge, Mario. 1997 [1980]. *Ciencia, técnica y desarrollo*. Buenos Aires: Sudamericana.

Bunge, Mario. 1998. *Social science under debate*. Toronto: University of Toronto.

Bunge, Mario. 2003. *Emergence and convergence*, Toronto: University of Toronto Press.

Bunge, Mario. 2009. The failed theory behind the 2008 economic crisis. In *Raymond Boudon: A life in sociology*, vol. 1, ed. M. Cherkaoui and P. Hamilton, 127–142.

Bunge, Mario. 2010. *Matter and mind*, Dordrecht: Springer.

Chomsky, Noam. 2009. *The mysteries of nature*, Journal of Philosophy 106: 167 – 200.

De la Losa, Pedro. 2000. *Espectro de Demócrito: Atomismo, disidencia y libertad de pensar en los orígenes de la ciencia moderna*. Madrid: Ediciones del Serbal.

Dragonetti, Carmen, and Fernando Tola. 2004. *On the myth of the opposition between Indian thought and western philosophy*. Hildesheim: Georg Olms.

Graham, Loren R. 1981. *Between science and values*. New York: Columbia University Press.
Lalande, Andre. 1938. *Vocabulaire technique et critique de la philosophie*, vol. 3. Paris: Alcan.
Latour, Bruno, and Stephen Woolgar. 1979. *Laboratory life: The social construction of scientific facts*. London/Beverly Hills: Sage.
Merton, Robert K. 1975. *The sociology of science*. Chicago: University of Chicago Press.
Needham, Joseph. 1956. *Science and civilization in China*, Cambridge: Cambridge University Press.
OECD. 2008. *Growing unequal? income distribution and poverty in OECD Countries.*
Panasiuk, Ryszard and Leszek Nowak, compils. 1998. *Marx's theories today*. Amsterdam-Atlanta: Rodopi.
Renen, Ernest. 1949. Averroès *et l'*averroïsme. Oeuvres complètes, vol. III. Paris: Coleman-Levy.
Soros, George. 1998. *The crisis of global capitalism*, New York: Public Affairs.
Stove, David. 1982. *Popper and after: Four modern irrationalists*. Oxford: Pergamon.
Trigger, Bruce G. 2006. *A history of archaeological thought*, Cambridge: Cambridge University Press.
Turiel, Elliot. 2010. Snap judgment? Not so fast: Thought, reasoning, and choice as psychological realities. *Human Development* 53: 105–109.
Wilczek, Frank. 2008. *The lightness of being*. New York: Basic Books.

Chapter 3
Systemics and Materialism

In the previous chapters it was claimed that one of the peculiarities of modern science since its birth five centuries ago is that it looks for systems—that is, complex objects with global or emergent properties. This attitude leads to analyzing wholes, as well as to placing them in their contexts.

*In all the fields of knowledge and action there is, with increasing frequency, talk of systemic problems and of the systemic approach to them. And yet most philosophers, with the exception of Holbach (*1770*) in mid-Enlightenment, have overlooked the systemic approach. So much so, that none of the standard philosophical dictionaries includes the entries 'system', 'systemic approach', and 'systemics'. What is more, none of them intimates that there is a connection between systemics and materialism, because the description of a concrete system involves specifying its mechanisms, which are material processes. These omissions suggest devoting a whole chapter to these concepts. In the following we shall endeavor to show the relevance of systemics to the whole of intellectual culture. We will begin with an increasingly serious social problem.*

3.1 The Housing Problem: A Component of a Ten-Dimensional Problem

Everyone agrees that housing is as basic a need as security, nourishment, and work. And yet, the housing problem is one of the worst global social issues: several billion people lack adequate housing. And, far from being solved, this problem is getting worse, because population and poverty are increasing almost everywhere. Let us see how individualists and systemists tackle it.

Individualists adopt a sectoral viewpoint: they believe that the housing problem can be resolved by just designing and building low cost dwellings. For example, Vijay Govindarajan and Christian Sarkar (2011), respectively a business professor and a consultant, have just proposed such a solution in the *Harvard Business*

M. Bunge, *Evaluating Philosophies*, Boston Studies in the Philosophy of Science 295,
DOI 10.1007/978-94-007-4408-0_3,

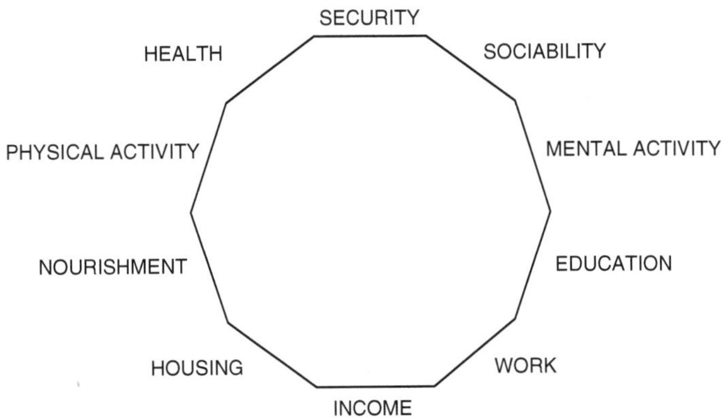

Fig. 3.1 The decagon of the satisfied (or at least bearable) life. When sides of the decagon are missing or too short, poverty results

Review with the plaudit of other prestigious periodicals. In the title the authors promise $300 *houses*, but in the text they admit that what they have designed are *sheds*, little better than tents, to be used in emergencies. They include a computer outlet but no sanitary appliances, hence no connection with the sewage and running water networks. (Why not go back to the ancient Roman columbaria, holes in a wall and accessible through hand ladders, which would be even cheaper?) Those business experts ignore city planning, invented and practiced 25 centuries ago by the ancient Greeks. They also ignore the many experiences in the Third World shantytowns. Here is one of them.

My great-grand-daughter Camila helped her father and his partners in an engineering firm build some low-cost houses—not mere sheds—in a Mexican "lost town". The owners were happy to move into genuine if modest houses.

But when Camila visited them a year later, she found out that the roof tiles, doors, windows, closets, and even sanitary appliances had gone. Vandalism? Of course not: the marginals are solidary: they survive by forming part of informal mutual help societies (Lomnitz 1977). What had gone wrong? What had happened is that the dwellers had sold those things to buy food. As the Mexican saying goes, hunger is a bitch.

Moral: The housing problem must be tackled systemically, not sectorally, because it is a component of the poverty problem, which is a whole package of system of problems. Other components of the poverty problem are occupation, the education required to get a job, the good health that must be enjoyed to hold a job, and the security needed to stay alive.

In other words, the problem of the satisfied, or at least bearable, life has at least ten interdependent components. Hence for life to become hard or even impossible, it suffices that one of them fails. See Fig. 3.1.

3.2 Approach

Routine problems are usually handled applying known rules. By contrast, radically new problems call for new studies, and every study involves some approach or point of view. Besides, every approach to a research problem involves philosophical hypotheses concerning the nature of the objects in question, as well as about the best way to study or manipulate them. For example, we need to know whether the object of interest is real or ideal, natural or artificial, as well as whether it can be subjected to calculation or experiment.

The simplest and most common approach is the sectoral one, which consists in isolating the object of study or action of whatever is not obviously related to it. This approach may be effective whenever one of the components or aspects of the item in question is far more salient than the others, but it will fail otherwise. In the vast majority of cases it will be best to assume that the object is not simple, and that it is not isolated either, so that it must be approached systemically.

The general concept of an approach may be analyzed as a body of background knowledge together with a set of *problems*, or problematics, a set of aims, and a set of methods or problem-solving techniques. In short, $A = \langle B, P, A, M \rangle$. In the case of the systemic approach, B = A general theory of systems, P = Any problems concerning systems, A = Understanding or management of systems of some kind, and M = Whatever regular and time-tested procedures will help attain the given aims.

A peculiarity of the systems approach is that it joins analysis (or top-down procedures) with synthesis (or bottom-up procedures) in addition to tackling same-level relations:

$$\begin{array}{ccccc} & \text{Whole} & \rightarrow & \text{Whole} & \\ \textit{Analysis} & \downarrow & & \uparrow & \textit{Synthesis} \\ & \text{Parts} & \rightarrow & \text{Parts} & \end{array}$$

Let us examine a few examples drawn from several disciplines.

3.3 Preliminary Examples

The ancient and medieval students of the human body, from the Egyptian embalmers to the medical charlatans derided by Molière, only saw mutually independent organs; moreover, they saw their patients detached from their social milieu. Such narrow focus is suitable for treating an ingrown toenail or an indigestion, but it is bound to fail when the sicknesss affects several interlocked organs or a social group threatened by a famine or an epidemic. Contemporary medicine, by contrast, is systemic, for it handles systems and supersystems of organs, as well as patients-in-their-environment, and even whole social groups.

Caution: Let us not confuse systemism with holism, a philosophy that is obscurantist because it rejects analysis. For example, so-called holistic medicine makes no use of any clinical analyses, and it does not put its conjectures to the test. Nor is holistic medicine interested in social medicine. In particular, it ignores that public health measures, such as the free supply of drinkable water, city-wide sewage, vaccination, and the control of environmental pollution have been more effective than all the pharmaceutical drugs put together (McKeown 1979). Epidemiology has thus invalidated the individualist recipe.

Psychiatry is of particular interest to philosophers For example, a classical psychiatrist might have assumed that a certain mental disorder was an abnormality of the immaterial soul as a whole. By contrast, the contemporary psychiatrist may suspect that it may consist in an abnormality of some brain subsystem provoked or worsened by a social circumstance. Thus the psychiatrist has adopted the systemic approach, perhaps unwittingly.

Law is another case of public interest. In this case too one may discern two trends: one of them detaches law from the other social studies, whereas the other looks at the legal system and the study of law through the socio-historical lens. A classical example of the sectoral approach is Hans Kelsen's "pure theory of law". The purity in question consists in the claim that law and legal practice are or ought to be amoral and apolitical.

This methodological requirement is unrealistic because almost all of the legal categories come from philosophy or the social sciences. Think of the concepts of good, harm, right, obligation, positive law, norm, proof, family, society, state, crime, or punishment: all of them are borrowed from disciplines other than the law. Besides, every legal system or code that does not fit some society is bound to be ignored or violated. Were it not so, no legal progress would be possible: we would still be practicing slavery or serfdom, obeying absolute monarchs, and so on.

Since *lex filia temporis*, the "pure" theory of law is just as false as the idea that there is such a thing as natural law—positive law being in fact typical social artifacts. Besides, because it equates justice with legality, legal positivism serves the *status quo*. (By contrast, natural law serves it only if conjoined with the nativist thesis that one is born either free or slave; if conjoined with Rousseau's thesis, that man is born free, then natural law can be used to fight slavery).

It was no coincidence that legal positivism was the official legal philosophy in both Nazi Germany and the Soviet Union. A true theory of law will be "impure" or interdisciplinary, ontologically as well as epistemologically. The former because it will assume society is a system of social systems; and the latter because it will make use of all the social sciences and technologies, as well as of practical philosophy—value theory, action theory, ethics, and political philosophy. See Chap. 10.

So far we have used tacitly the concepts of systemic and sectoral approach. In the following we shall review the salient traits of systemics, after which we shall review some recent contributions, from physics to public management.

3.4 Systemic Approach and Theory

Systemics may be defined as the systemic approach together with general systems theory. The former consists in adopting a viewpoint both comprehensive and analytic: in checking whether complex objects are systems rather than either simple individuals or amorphous collections in a vacuum. And a general theory of systems studies the common traits of systems of all kinds: natural and artificial, concrete and conceptual.

Every general theory of systems may be conceived of or applied in either of the following ways:

ontological, or representative de real systems such as families,
epistemological, or representative of knowledge systems such as theories,
praxiological, or tool for managing real systems such as businesses.

The simplest and most general model of a system σ consists in the ordered quandruple

$$\mu(\sigma) = < C(\sigma), E(\sigma), S(\sigma), M(\sigma) >$$

where
$C(\sigma)$ = *Composition* or set of parts (on a given level),
$E(\sigma)$ = *Environment*,
$S(\sigma)$ = *Structure* or set of relations among the parts,
$M(\sigma)$ = *Mechanism* or peculiar process in the system,

If the system of interest is conceptual, such as a legal code or a theory, or semiotic, such a text or a diagram, then the last component of $\mu(\sigma)$ is the empty set, and the first three components will be conceptual or semiotic. The mechanism of a concrete or material) system is the process that makes it what it is, such as metabolism in a cell, and learning in a school.

For example, a model of a learned international systemics society σ will describe in detail the following aspects of σ:

$C(\sigma)$ = The set of national systemics societies,
$E(\sigma)$ = The set of national higher cultures,
$S(\sigma)$ = The interactions among the members of $C(\sigma)$,
$M(\sigma)$ = The management of σ and the meetings convened by σ to advance systemics.

Let us next glimpse at a few recent developments that confirm the systemic theses.

3.5 Natural Sciences

We shall recall only a couple of recent triumphs of systemics in the natural sciences, one in physics and the other in biology. One of the peculiarities of quantum physics is the persistent entanglement of the components of a system even after they have separated in space. This is a quantum peculiarity, because according to classical physics all bonds weaken with distance. The conservation of systemicity allows one to predict the properties of one of the components of a binary compound from a measurement on the other.

This trait of the quantum theory, pointed out by Einstein and Schrödinger in 1935, was experimentally confirmed half a century later. Recently it was corroborated observing two components of a system, one in each of two Canary islands 140 km apart. There is no doubt now that *once a system, always a system.* Like all great scientific achievements, this result is counter-intuitive.

Systemic thought began in the life sciences with William Harvey's work on the cardiovascular system. Since then several other biosystems have been discovered: digestive, nervous, endocrine, immune, the plant communities and the ecosystems. As a matter of fact the unit of life, the cell, is a system, and every organ is a system of cells, so that the life sciences study systems on several levels.

Ecology has been systemic from birth, since it has always dealt with interacting organisms. In recent years an increasing number of ecologists have been studying the practical problem of reconciling the so-called nature conservation with human needs. The radical environmentalists such as the Greenpeace activists, will have to face the problem of meeting human needs. For instance, how can the Amazonian forest be protected along with their human dwellers?

Systems biology, which studies the systems in individual organisms, is becoming fashionable after a long period of dominance of radical reductionism. For example, unlike classical genetics, which focused on individual phenotypical traits, such as eye color and height, systemic genetics tackles phenotypical properties that come in bundles—Darwin's "mysterious correlations".

Another recent triumph of the systemic approach in the life sciences is the birth of *evo-devo,* short for evolutionary developmental biology. This merger of two formerly separate disciplines became unavoidable the moment it was realized that evolutionary novelties emerge in the course of the development of the individual organism. In turn, this discovery could not be accepted as long as population thinking prevailed, since in this perspective individuals are but parts of populations, and evolution would be peculiar to populations and, moreover, it would consist exclusively in certain global traits, namely changes in the frequencies of certain genes in populations. It is now well understood that biological evolution happens on all levels: genome, organ, organism, population, and ecosystem. Furthermore, biologists reason from part to whole and back. This is only logical, for, by definition, there is no part without whole, and no whole without parts.

Further mergers are in the offing, among them those of genomics with proteomics, and of ecology with evolutionary biology. The former fusion is needed because genes control protein synthesis, and in turn enzymes (a kind of proteins) split genes. And the second merger is desirable because genetic changes makes some ecological niches more inhabitable than others.

In short, in the life sciences systems thinking is quickly replacing both holism, obvious in vitalism, and individualism, inherent in certain popularizations of molecular biology.

3.6 Social Sciences

Although all people gather in systems of various kinds and scales, the systemic approach to society is still minoritarian: most scholars focus either on "great men" or on entire "cultures". The Islamic scholar Ibn Khaldün (1377?) is likely to have been the earliest systemic social scientist: the first to treat all the aspects of the entire Islamic world of his own time. The next breakthrough in the systemic study of society was that of the French *Annales* historiographical school, whose best-known work was Fernand Braudel's monumental study of the Mediterranean basin in the age of Philip II (1949). He and his colleagues practiced systemism without proclaiming it. Immanuel Wallerstein (1974) was the first to tackle the world system, and to note that a nation's affairs depends critically upon its place, central or peripheral, in this system.

The systemic approach has been gaining ground in anthropology and archaeology. Whereas formerly kinship relations, work, or beliefs and ceremonies were privileged, in recent years anthropology has become the total social science: the one that studies all the aspects of life in the primitive, archaic and modern "cultures" (societies). Thus Bronislaw Malinowski's "functionalist" (systemic) principle, that all the traits of a society are interdependent, is being put into practice in contemporary anthropology.

Archaeology has followed a similar path. Indeed, for the past century archaeologists the world over have been trying to reconstruct, both mentally and in museums, the way people of old used to live, in particular what they did to make a living, be sheltered, and defend themselves. His is how archaeology, which had been just antiquary, became a social science. And this is why it poses so many interesting philosophical problems, as Marvin Harris (1968) and Bruce (2003) showed.

Malinowski's systemic principle is ignored by economism, both on the left and on the right. In particular, standard economic theory is individualistic both ontologically and epistemologically. Indeed, this theory focuses on businessmen and consumers, and ignores the relevance or even the existence of social systems, such as the state, as well as the need to learn from anthropologists, sociologists, and other students of society. Unsurprisingly, that theory has turned out to be a monumental flop. Because of its emphasis on equilibrium, it has proved utterly unable to account for economic disequilibria and innovation. Worse, it has inspired the neo-liberal policies that have powerfully contributed to the recent economic crises. (More on this in Chap. 8.)

The best noticed among the recent triumphs of systemics in the social studies is the emergence of socioeconomics, the merger of sociology and economics. This fusion has been particularly useful in the study of socioeconomic inequalities, especially in the design of indicators of such inequalities, as well as in the design of socioeconomic development policies.

To sum up, although individualism is still dominant in sociology and economics, systemism has been advancing in the other social sciences. (More in Bunge 1998; Wan 2011).

3.7 Biosocial Sciences

Psychology, which until recently was regarded as an independent discipline, is now being increasingly treated either as a synthesis of psychology, linguistics and information technology, or as a product of the fusion of neuroscience with psychology and a fragment of sociology. The first of the two mergers, called 'cognitive science', was the cheapest and the one that attracts folks educated in the humanities or in the information sciences, where they were not required to think in neural terms, let alone handle squishy brains.

This is also the synthesis requiring the least intellectual effort, since it only involves the allegedly universal concepts of computation and information. By proceeding in this manner one glosses over the differences between computer and brain, passion and reason, search and routine, law of nature and algorithm. What is gained in simplicity is lost in understanding: one fails to understand why the computer is incapable of starting anything, why it cannot invent, and why it is insensitive to the social milieu.

The so-called cognitive science ignores what goes in living brains, those studied by the neurosciences. And what these find are not algorithms but natural laws such as the so-called Hebb rules. It should not be surprising that the brainless approach, which presupposes mind-body dualism, has not led to any important discoveries. Idealist systemism may be fertile in the study of conceptual and semiotic systems, but it is bound to be barren in the case of concrete entities, such as us.

Cognitive science has no use for the social aspect either. But the findings of social psychologists cannot be ignored: there is no doubt that learning is done in a social milieu and that is the better, the richer that milieu. This realization has led cognitive neuroscientists to enlarge their science to social cognitive and affective neuroscience—the widest of biosciences. This science has made sensational discoveries with the help of imaging techniques such as fMRI, which allow researchers to locate mental functions—thus refuting Kant's contention that the mind exists in time but not in space. And the mere existence of this bioscience refutes once more the neo-Kantian thesis that there is a an abyss between the natural and the social (or cultural) sciences.

The new imaging techniques have led to joining two ideas that used to be regarded as mutually inconsistent: localizationism and holism. We now know that the brain is neither a collection of ultraspecialized modules (the so-called Swiss Army penknife model) nor a homogenous body that acts like an undifferentiated blob. Just like other subsystems of the body, the human brain combines specialization with coordination.

The philosophical dividends of social cognitive neuroscience are obvious: It has confirmed the psychoneural identity hypothesis—a piece of ontology—, as well as the research strategy involving the convergence of disciplines that used to be cultivated in mutual independence: neuroscience, psychology, and sociology.

Another important discovery has been that the nervous system is intimately connected with the endocrine and immune systems, which are in turn strongly influenced by the social milieu. This is why one may speak of the most comprehensive of syntheses: psycho-neuro-indocrino-immuno-sociology. Applied scientists will add one more link: pharmacology. The psychotropic drugs they design would be ineffective unless the first chainlike system were real.

In sum, the process we have just summarized has involved a system of organs and the corresponding system of basic sciences, which in turn feeds the system of technologies, which feed the industry. The four systems concerned are embedded in society and are linked thus:

Sciences → Technologies → Industries
Study ↘ ↙ Treat
Human body

3.8 Technologies

Engineers have always adopted the systemic approach, since all artifacts are systems. However, they have usually circumscribed the systemic approach to their creations, forgetting sometimes that they have users, and that artifacts, their users and the environment constitute a supersystem. An extreme example of user neglect was the Millennium Bridge over the river Thames. When this beautiful artifact was inaugurated in the year 2000, it started to oscillate horizontally as the first pedestrians rushed onto it. The designer, the eminent engineer Lord Norman Foster, had forgotten to include in the equations the pressure that a person who, so as not to lose balance, exerts on the floor that moves in the opposite direction. The equations were easily corrected and the bridge was repaired at great cost and with some loss of elegance. Presumably, this fiasco would have been averted if an expert in human engineering had been consulted.

Social engineering made a great step when the UN Development Program replaced the GDP with the HDI (Human Development Index) as a reliable measure of human development. This innovation involved a replacement of the sectoral with the systemic approach. Still, the new index is not sufficiently systemic because it does not include income distribution, political participation, or environmental sustainability.

Finally, let us recall the most urgent of all tasks of all designers of socioeconomic policies: how to feed the world population in the next decades. This problem is getting rapidly more intractable because the world population keeps growing while the cultivated area keeps decreasing, the water tables keep descending, the amount and fertility of the soil keep decreasing, freak weather is becoming more frequent, and so on. Obviously, wiggling a single variable won't do.

True, the Green Revolution has enormously increased agricultural yield, but at the price of a great increase in irrigation and fertilization, as well as a higher price for seeds, increasing soil erosion and in the concentration of landed property, and the resulting impoverishment of peasants. In short, grain has become more plentiful but also more expensive.

Another source of the increase in the price of grain is that it is being subsidized for its production of either animal feed or biofuel. All of this has decreased the production of food, which is in turn fueling the inflation that nearly all nations are suffering. In short, rapid agricultural progress is causing a planetary food crisis along with the enrichment of comparatively small social groups. The political lesson is clear and well known : technology without social control can be catastrophic.

The task is as obvious as it is huge: To design and put into practice new demographic, agricultural, and business practices that may feed seven billion humans instead of benefiting almost exclusively big agrobusiness, and seed and insecticide producers. In sum, preparing for the coming world food shortage calls for a systemic approach on a planetary scale. But I am afraid that this task is beyond the pertinent authorities, in particular the World Bank, who are likely to keep distributing alms nation by nation—typical of the sectoral approach.

3.9 The Knowledge System

It has been known since Aristotle that the various fields of knowledge are closely related to one another. In particular, they have been imagined as branches of a tree, steps in a ladder, and petals of a rose. For example, all modern biologists know that biology, though peculiar, uses a lot of chemistry, which in turn is based on physics. (We shall come back to this in Chap. 18.)

However, most specialists are isolationists. Hans Kelsen's "pure theory of Law" is an example of the grotesque result that disciplinary isolationism can lead. According to this legal philosophy, a version of legal positivism, legal theory deals only with positive law, and is or ought to be free from both ethical and political considerations. According to this conservative view, the jurist should serve the status quo, however perverse, instead of proposing legal reforma. But, of course, for better or for worse, there have always been "activist" judges and progressive jurists, such as Roscoe Pound and the Scandinavian realists. All of them have known that law, far from being autonomous, is at the very center of the intersection of politics with morals and social science, and that it may advance or regress thanks to such "impurities".

Only logic has the privilege of not (logically) presupposing any other disciplines, even though in ancient Greece it was the twin of law and developed in close partnership with mathematics. And mathematicians presuppose only logic: mathematicians can afford to ignore the sciences of reality because they do not touch them except at the request of scientists or engineers. Mathematics, which nowadays includes logic, is the universal research tool in all fields, so that it must be placed at the very center of the rosette of the disciplines. In fact, it is used even in fields that, like philosophy, history, and psychology, seemed to lie forever beyond exactness.

The reason that the sciences of reality overlap partially is that the universe is one: it is we who tend to quarter it to investigate it in detail, and then forget that we have started with a single thing. Analogously, the supermarket dependent can afford to ignore what the village butcher knows by heart: that the starting point is the whole animal.

This brings to mind the economists who believe that they study and control the economy as if it were independent from nature, politics, and culture. This sectoral approach of theirs is one of the reasons for their utter failure; the other main reason is their assumption that the market is always in equilibrium or very near it, so that only a gentle nudge will suffice to fix it. But neither crises nor criticisms have persuaded the schools of economics that they need change ppdate the teachings of Jevons, Walras and Menger except in mathematical details.

The sectoral approach can bring small doses of knowledge, but it cannot tackle systems of any kind, whence it is bound to hinder the study and control of reality which is the system of all sytems

3.10 Philosophical Systems

Real life problems come in clusters rather than one by one. This is why useful philosophies are systematic, and also why philosophical analyses are best performed within systems where every idea is supported by other ideas. Take again the case of law. The lawyers or judges called to hear a criminal case will start by finding out whether it is true that a crime was committed: thus they presuppose de correspondence theory of truth. They also assume the distinction between objective facts and data about them: they are not social constructivists.

Furthermore, far from trusting witnesses or their intuition, the jurists will look for evidence to confirm or infirm the accusation of swindle. They will also make tacit use of the ethical definition of crime as unjustified harm, or alternatively as breach of a moral norm entrenched in positive law. Thus jurists do not circumscribe themselves to analyzing and applying codes of law: they debate with themselves and with colleagues, and use ontological, epistemological, methodological, and ethical theses. The reason for such multitude of sources is that ideas do no come in isolation but in bundles or systems—particularly if they are about the real world.

However, typically the so-called analytic philosophers reject the idea of a philosophical system, and prefer to detach ideas from each other and from their contexts: they prefer sawdust to logs, and logs to forests. This is why the philosophical censor of the oldest university press in the world gave me this piece of advice when he rejected my book *Matter and Mind*: "Pick a circumscribed subject and engage with other contemporary philosophers who have written about it." No doubt, this is a good recipe for writing a term paper for a standard philosophy course, or for engaging in scholasticism, but not for doing original work. Original workers, in any field, will certainly start working on a limited research project, even if they tackle a big problem. But they cannot anticipate where their research will lead: if they could, they would accomplish a routine task, not an original investigation.

We must of course distinguish the various branches of philosophy, but if we cut them from the trunk they will wither and die. For example, even though epistemology differs from ontology, it gains when treated as the ontology of knowing, which in turn is enriched if it treats knowing as a brain process, for then epistemology can learn from cognitive psychology. Incidentally, epistemological systemism is called *coherentism*, and it is usually regarded as the alternative to foundationalism. I suggest that the two approaches are mutually complementary: as has been know since Euclid, if circularity is to be avoided, some of the hypotheses in a theoretical system must be postulated.

Here is another example: Ethics is empty and barren when culivated in a vacuum, but it can be enriched and utilized when based on value theory and the social sciences, and when thought along with action theory and political philosophy. *Divide and conquer* holds in politics. In matters of knowledge, the ticket is *Unite and Conquer.*

3.11 Concluding Remarks

The systemic approach has gained much ground, especially in mathematics, the natural sciences, and engineering, since its birth around 1600. But it is still backward in the social sciences, even though all of these study but social systems, something they will ignore only at their risk. If only for the sake of social technology, which faces daunting systemic problems, it is desirable that the social studies overcome the individualist/holist dualism. It is also desirable that the same approach be adopted once again in philosophy since, by definition, every big philosophical problem has logical, ontological, epistemological, value-theoretic and perhaps also praxiological aspects.

References

Braudel, Fernand. 1972 [1949]. *The mediterranean and the mediterranean world in the age of Philip II*, 2 vols. New York: Harper & Row.

Bunge, Mario. 1979. *Treatise on basic philosophy*, vol. 4, *A world of systems*. Dordrecht, Boston, Lancaster: Reidel

Bunge, Mario. 1998. *Social science under debate*, Toronto: University of Toronto Press.
Bunge, Mario. 2003. *Emergence and convergence*. Toronto: University of Toronto Press.
Govindarajan, Vijay & Christian, Sarkar. 2011. *The $300 house*, Harvard Business Review, October.
Harris, Marvin. 1968. *The rise of anthropological theory*. New York: Crowell.
Holbach, Paul-Henry Thiry. 1770. *Système de la nature*, 2 vols. Repr. Hiildesheim: Gorg Olms, 1966.
Holbach, Paul-Henri Thiry Baron d', 1773. *Système Social*, 3 vols. Repr. Hildesheim: Georg Olms, 1969.
Ibn Kaldûn. 1976 [1377?]. *The Muqaddimah*, 3 vols, 2nd ed. Trans. F. Rosenthal. London: Routledge & Kegan Paul.
Lomnitz, Larissa. 1977. *Networks and marginality: Life in a Mexican Shantytown*. San Francisco: Academic.
McKeown, Thomas. 1979. *The role of medicine*, 2nd ed. Princeton: Princeton University Press.
Trigger, Bruce G. 2003. *Artifacts & ideas*. New Brunswick: Transaction.
Wallerstein, Immanuel. 1974. *The modern world-system*, 3 vols. New York: Academic Press.
Wan, Poe Yu-tze. 2011. *Reframing the social*. Farnham/Burlington: Ashgate.

Part II
Philosophy in Action

Chapter 4
Technoscience?

The daily press and most of its readers tend to confuse science with technology. For example, the Science section of most newspapers reports news in engineering and medicine. By contrast, Science, Nature and the other scientific periodicals publish almost exclusively papers in disinterested research.

Conservative governments either neglect science altogether or tend to support only research projects that promise practical results. These promises are usually empty, because only small improvements are foreseeable and can be designed on demand. This is how such governments cut the innovation stream as well as an unknowable number of scientific careers.

The confusion between science and technology has recently been given a new name: technoscience. This newcomer has spread across Internet, as well as among postmodern scribes and improvised sociologists of science. It is therefore necessary to alert against that confusion, if only to allay the fear of science. Rocket engineering is dangerous, but rocket science is harmless because it does not exist.

4.1 Discovery and Invention

The confusion between science and technology is quite old. Francis Bacon, the great modern philosopher, was one of the first to incur it, and in arguing for scientific research because of its alleged practical fruits. However, four centuries ago Bacon saw science as detailed description of the visible world by a small team of meticulous, industrious and unimaginative observers, rather than as the great thought adventure it was for Galileo, Harvey, Boyle, and their followers.

The difference between science and technology ought to be obvious, and so it is to most of their practitioners if only because they are studied in different faculties: whereas scientists study reality, technologists seek ways of changing it. In other words, whereas scientists work for the sake of learning, technologists work for the sake of doing.

M. Bunge, *Evaluating Philosophies*, Boston Studies in the Philosophy of Science 295,
DOI 10.1007/978-94-007-4408-0_4,

For example, whereas nuclear physicists study atomic nuclei, nuclear engineers employ some (very few) findings of the former in designing nuclear reactors or nuclear bombs; biologists study organisms, and biomedical investigators use some of the former's findings to design therapies; whereas psychologists study learning, education scientists study learning to perfect teaching methods; and whereas sociologists study business firms to find out how they are organized, management scientists study them to improve their performance.

Henry Petroski (2010), the well-known engineering scholar, gives a simple and eloquent example: Whereas the astronomer knows how to detect meteorites moving towards our planet, the engineer seeks to deviate their trajectories. Likewise the Ayatollah Khomeini, the first leader of the Islamic Revolution in Iran, knew the difference between the two fields. When the journalist Oriana Fallaci asked him what he would do with them, he answered unhesitatingly: Technology will be allowed because it is useful, but science will be banned because it is dangerous.

Another aspect of the question that ought to be clear is that some sciences have no technological "translations". Examples: cosmology, archaeology, paleoanthropology, ancient historiography, linguistics, musicology, philology, and mathematical economics.

Last, but not least, there is the socioeconomic aspect: whereas scientists produce public goods, technologists produce private goods. Indeed, the inventions with possible practical uses are patentable and therefore marketable, whereas the scientific ones are not. (Think of mathematical theorems or archaeological findings).

It is not for nothing that Philip Mirowski's (2011) eloquent attack on the commercialization of American science is directed at the applied sciences such as the biomedical ones, but only touches lightly on the basic ones, such as particle physics, with regard to cuts on public funds for them.

4.2 Primacy of Praxis?

Despite the marked differences between basic science and technology, many intellectuals keep conflating them. Some out of sheer ignorance, and others because they wish to shift our attention from contemplation to action. The latter is the case of the pragmatist philosophers. The exaltation of praxis by Heidegger and other existentialists has a different source: they praise blind action and despise knowledge.

In any event, the said conflation is ontologically wrong because it ignores that all technologies handle artificial things, such as tools, buildings, farms, genetically modified organisms, shops, schools, hospitals, and courts of law: all of these systems are man-made. Aristotle had already noted the differences between the artificial and the natural, the made and the found. The early modern thinkers overlooked these differences, perhaps because they were intent on building a naturalistic worldview.

The confusion in question is also epistemologically mistaken because, whereas scientists try to understand natural or social entities that they find or discover, the innovative technologist attempts to design things that do not yet exist, such as new cultivars or farm animals, means of communication or transportation, cooking recipes or pieces of legislation.

Lastly, the confusion of scientists with technologists is also axiologically and praxiologically wrong, as is realized by watching what they do. Indeed, whereas scientists do not evaluate their objects of study, technologists are only interested in things or processes of possible practical utility. This is why some companies subsidize pharmacology laboratories, but not astronomical observatories. In other words, whereas technology and applied science are utilitarian, basic science is disinterested.

Disinterestedness is one the peculiarities of basic research. Others, as Robert Merton (1973) famously noted, are universality and communism (the sharing of knowledge). The new breed of sociologists of science—such as David Bloor, Harry Collins, Steve Fuller, Bruno Latour, and Andrew Pickering—exhibit alleged counterexamples that would disprove Merton's thesis. But they take all their cases from the intersection of science and technology. A perusal of the main journals in nuclear and atomic physics, the usual suspects, will show that the overwhelming number of papers in these fields show no technological prospects: they are products of the combination of the two main drives of pure scientists: curiosity and peer recognition.

4.3 Consequences of the Confusión

In short, science and technology are ontologically, epistemologically, axiologically and praxiologically quite different from one another. The confusion between them is one of the may philosophical errors spread by the media since their inception. This error prevents most people from noticing the trench separating both fields, and thus hinders bridging it. It also leads to the futile attempt to transmute basic scientists into technologists or even entrepreneurs.

The confusion in question has led to the commmodification of certain scientific findings. A recent egregious example is constituted by the lawsuits generated by attempts to patent the genes involved in certain medical treatments. Every time such a lawsuit is won by a company, the most basic right of all, to the ownership of one's own body, is violated. David Koepsell (2009) has explored this matter in his chilling *Who owns you?*

The conflation of science with technology underlies myopic cultural policies, in particular cutbacks in scientific budgets. Here is where the philosopher can intervene, to prevent that the goose of the golden eggs be killed. Indeed, he can teach cultural policy makers that basic science is the fountain of modern civilization, since it feeds technology, which in turn feeds industry and the state:

Industry
↗
Science → *Technology*
↘
State

4.4 "Translation" of Science into Industry via Technology

The arrows in the preceding diagram symbolize extremely complex processes that are the object of "translational" research. This activity puts basic scientists in touch with technologists, and it has its own journals in the case of medicine. For example, it is not obvious how to "translate" a biochemical discovery into a therapy, that is, how to go from the laboratory to the factory, and from the latter to the clinic.

The large-scale manufacture of a molecule analyzed or synthesized in the laboratory is likely to present new challenges to the chemical engineer and to the factory manager. Whereas some of these challenges are technological, others are economic, and still others are moral. For example, it often happens that the industrial process is highly contaminating, which poses the problem of containing it or even of replacing it with a "green" reaction having the same products but less polluting reagents.

It may also happen that patenting a new powerful pharmaceutical product allows the industry to sell it at a price inaccessible to most patients. The industry claims that is due to the high costs of the research and "translation" processes. But the truth is that most of the former take place in universities, and most of the latter in government laboratories.

For example, the main mission of the National Institutes of Health, of the US government, is to "translate" basic science into industry at the cost of about 32 billion dollars per year. Hence the American taxpayer pays thrice for the research involved in the production of any pill: to the pharmacy, to the Internal Revenue Service, and to the college that educates his children. This public contribution is so large, that some pharmaceutical giants are dismantling their laboratories.

4.5 Authentic Technosciences

To sum up, there are marked differences between basic research and technology. Should we conclude that there is no such thing as technoscience? Certainly not, since there are multidisciplinary fields, such as engineering, medicine, agronomics, economics, and the law, that are cultivated by scientists that help design new things, as well as by technologists that study physical, chemical, biological or social artifacts in a scientific manner.

For example, whereas some engineers invent new materials. others seek to use them, and still others study their physical or chemical properties. There are biomedical investigators who study sicknesses but do not see patients; and, whereas some study the mechanisms underlying therapies, others study their effects at the organismal level. Some epidemiologists, medical sociologists and economists engage in social engineering. Whereas most lawyers mediate in conflicts of interests, some jurists study legal codes. And whereas some criminologists study criminogenesis, others seek to decrease the criminality rate.

All of these students and professionals are technoscientists to the extent that they use the scientific method and employ scientific findings to study, control or design artificial processes or things, such as machines, organisms, medical treatments, schools, businesses, and so on. However, with few exceptions these individuals have been mainly technologists. Only very few—Archimedes, Galileo, August Liebig, Joseph Henry, Louis Pasteur, Nikola Tesla, and Enrico Fermi—have done great technology as well as great science. It would be ridiculous to call Newton or Darwin technologists, or Marconi and Edison scientists.

The strongest motivation of the basic researcher is curiosity, not utility. For example, molecular biology was motivated by the wish to know the chemical composition and molecular structure of the hereditary material. And CERN, the European scientific concern, invested billions of dollars in building the Large Hadron Collider, whose function is to test theories such as the one that conjectures the existence of the Higgs boson. Nobody hopes to recover even one euro from that gigantic investment in knowledge for the sake of knowing. Aristotle would have appreciated this project, because he believed that research is driven by curiosity.

4.6 Conclusion

The neologism *technoscience* originated in a confusion, and it has consecrated the popular superstition of the mad scientist as well as the myopic science policies of backward politicians. Hence it should be use sparingly if at all.

Actually it is preferable to use the expression *applied science*, which denotes the scientific investigation that seeks results of *possible* practical utility, such as molecules that might be used in therapies, genetic modifications of cultivars, and programs of criminality control.

The following table summarizes the peculiarities of the three fields of knowledge involved in the preceding discussion

Basic science	*Applied science*	*Technology*
Disinterested study of reality	Investigation with possible practical utility	Design of artifacts

In conclusion, science and technology are quite different fields of intellectual culture. Still, detaching them would be just as wrong as confusing them, because some science—the more the better—gets translated into technology. For example, the design of medical treatments is based on a knowledge of disease mechanisms, and these can only be unveiled through basic research.

Moral: Distinguish science from technology and support them both while fostering their vigorous interaction.

References

Agassi, Joseph. 1985. *Technology: Philosophical and social aspects*. Dordrecht/Boston: D. Reidel.
Becker, Gary S. 1976. *The economic approach to human behavior*. Chicago: University of Chicago Press.
Bunge, Mario. 1985. *Treatise on basic philosophy, Part II*, vol. 7. Dordrecht/Boston: Reidel.
Koepsell, David. 2009. *Who owns you?* New York: Wiley-Blackwell.
Kranzberg, Melvin, ed. 1996. Toward a philosophy of technology. *Technology and Culture,* 7(3): 301–390.
Merton, Robert K. 1973. *Sociology of science*. Chicago: University of Chicago Press.
Mirowski, Philip. 2011. *Science-Mart: Privatizing American science*. Cambridge, MA: Harvard University Press.
Mitcham, Carl. 1994. *Thinking through technology*. Chicago: University of Chicago Press.
Ortega y Gasset, Jose. 1939. *Meditación de la técnica*. Buenos Aires: Espasa-Calpe.
Petroski, Henry. 2010. *The essential engineer*. New York: Alfred A. Knopf.
Quintanilla, Miguel Angel. 1988. *Tecnología: Un enfoque filosófico*. Madrid: Fundesco.
Susskind, Charles. 1973. *Understanding technology*. Baltimore: Johns Hopkins University Press.

Chapter 5
Climate and Logic

A whole battery of reliable indicators shows that our planet is warming up: the temperature is going up, the glaciers are melting, the Arctic ice cover is evaporating, the Antarctic is releasing icebergs, the snow cover is thinning almost everywhere, etc. The only climate skeptics are the corporations and economists more interested in today's gains than in the future of their own offspring.

Given the reality and danger of climate change, and the political clout of the deniers, it is important that climate scientists, environmental activists and policy makers wield valid arguments and propose well-grounded and feasible measures to control climate. Unfortunately, sometimes they incur elementary errors. One of them is the Kaya identity, used by the IPCC (International Panel on Climate Change) in its climate studies and recommendations. Let us examine this famous formula.

5.1 The Kaya Identity

So far, nearly all proposals to limit the negative effects of industry and transportation on the environment have boiled down to emission control. Now, any engineer knows that, to control a material system, one must vary its inputs, or driving factors, not its desirable outputs, for these are goals, not means. In particular, the technologist will focus on the so-called knob variables, that is, those that, unlike the solar and geological ones, can be effectively controlled by people.

Suppose now someone believed that environmental degradation is the Devil's work. If convinced also that the problem merits quantification, he might propose the following formula for the volume C of global carbon emission in terms of the number D of demons at work:

$$\mathrm{C} = \mathrm{D} \times (\mathrm{C} / \mathrm{D}) \tag{5.1}$$

People with a secular worldview would laugh, and logicians would object that, since (5.1) is nothing but the identity "$C = C$" in disguise: it cannot say anything

M. Bunge, *Evaluating Philosophies*, Boston Studies in the Philosophy of Science 295, DOI 10.1007/978-94-007-4408-0_5, © Springer Science+Business Media Dordrecht 2012

about reality even though it refers to a real item, namely C. Indeed, it only says that C is C—unobjectionable but empty.

The preceding story is intended to caution the economists interested in climate change. In 1993 the Japanese energy economist Yoichi Kaya proposed the so-called *Kaya identity*, that has been the centerpiece of both climate policy-makers and their critics. This formula is

$$C = P\times(GDP / P)\times(E / GDP)\times(C / E), \tag{5.2}$$

that shows the dependence of the volume C of carbon emissions upon population P, per capita gross domestic product GDP/N, energy intensity of the economy E/GDP, and emissions per unit energy C/E.

The above formula is unassailable: it holds in any weather under all climates. In fact, it is but a rewrite of the identity "$C = C$", which is the same as "Carbon is carbon." Hence it is impervious to empirical data. Consequently, that formula cannot serve as the basis of climate control policies, anymore than "Either you do it or you don't." Trying to force logic to tell us something about the world is almost as mistaken as rejecting it altogether, the way the post-moderns do.

Still, it is true that the volume of carbon emissions depends on the variables indicated above, and that all of them are knob variables, that is, variables that, up to a certain point, can be wiggled so as to change the output as desired. For example, P (population) can be controlled directly the way China does; or indirectly, through raising the standard of living, the way the West has been doing for the past century. And the values of last two variables in (5.2) can indeed be lowered through technological advances. So, it is worth looking for ways of transforming (5.2) from a tautology into a statement about reality. (In Sect. 19.2, we shall return to the difference between identity and equality.)

5.2 From Logic to Reality

One way of moving (5.2) from logic to reality is to include the savings rate σ characteristic of the society in question. The assumption is that σ is a handy indicator of spending habits, from saving part of the income to turning out the lights when leaving a room, to keep wearing old but still serviceable clothes.

As we know, savings are highest in Western Europe and Japan, and lowest in the USA and the developing countries. The parameter σ cannot be increased any further where people live on one dollar or two a day. But it can certainly be significantly increased wherever people have become used to wasting energy on air coolers, SUVs, and golf courses, as well as on throwaway manufactured articles.

In any event, σ is a cultural variable, not a technological one. Hence it can be altered by education. This is why the Swedes, who enjoy a significantly higher level of wellbeing than the Americans, and on the whole a colder climate, spend roughly half the energy per capita.

Another key variable is technological inefficiency. I suggest that a possible measure of such inefficiency for a whole nation is TMR, or total material requirement—the amount of natural resources required to generate one unit of economic activity (see Bringezu and Schütz 2001). Recent values of TMR are about 45 t per capita and year for Japan, 49 for the European Union, and 84 for the USA.

5.3 A New Formula

We tentatively propose the following formula for the volume of global carbon emission:

$$C = P.GDP.(1-\sigma).TMR, \tag{5.3}$$

Note that σ is negative wherever most people live on credit. And the equation is too simple to be true, but at least it is no phony, so that it might serve to start a serious discussion.

The practical lesson is obvious: To decrease *C*, we should favor limiting the growth of both population and GDP, as well as increasing the savings rate σ and decreasing the technological inefficiency TMR.

5.4 Conclusion

There is some talk of geoengineering to control the climate. An economist has even suggested covering the planet with myriads of small mirrors, to shield it from solar light—a project worthy of Jonathan Smith's Lagado Academy. In a frequently quoted paper, Galiana and Green (2010) hope that a "technology race" may end up by finding a technological fix—to implement the empty Kaya identity.

I suggest that, since climate is affected by deeply ingrained habits in all walks of life, climate control requires mobilizing humankind so as to alter them. We need to limit population growth, manage the non-renewable resources, exploit alternative energies, prevent wars, learn energy-saving habits, and significantly increase the efficiency of both production processes and artifacts. It might also help if economics students were to take Logic 101.

References

Bringezu, Stefan, and Helmut Schütz. 2001. *Total material requirement of the European Union.* Copenhagen: European Environment Agency.

Galiana, Isabel, and Christopher Greene. 2010. Let the global technology race begin. *Nature* 462(3): 570–571.

Chapter 6
Informatics: One or Multiple?

It is no secret that information technology (IT) is going through an identity crisis. A major subject of controversy among experts in this field has been whether it is a single discipline or a conglomerate of distinct fields joined by "the" concept of information. Opinion on this question seems to be split roughly in half. Moreover, the very core of such disciplines, namely the concept of an information system, is the object of unresolved dispute (see, e.g., Beynon-Davies 2010*). Let us take part in this debate from the systemic viewpoint expounded in Chap. 3.*

It is well known that there are a number of disciplines that claim to deal with information, among them computer science, communication engineering, and even genetics, linguistics, and library science. At first sight, these fields are one because they all focus on information. Moreover, many believe that Claude Shannon's communication theory is so general, that it is applicable everywhere, from physics to biology to social science. Let us see whether these claims are true.

6.1 From Information System to Communication System

Shannon's famous formula involving the probability of a signal has been tried in many fields of basic science, from physics to genetics to sociology, in addition to information technology. However, the word 'information' is polymorphous. In fact, it can mean knowledge, coded signal, quantity of order, the order of nucleotides in a DNA molecule, and more (Mahner and Bunge 1997). Moreover, a coded signal can be sharp, as 01, or smeared, as a quantum-theoretical mixture of 0 and 1. Given this great variety of information concepts, it is doubtful that there is *conceptual unity* among them. Consequently, it is also doubtful that a single theory, such as Shannon's, will fit all those different senses of 'information'.

What is clear is that Shannon's theory deals with the communication of messages regardless of the information they may carry. So much so, that it applies to messages of all kinds, even questions, orders, and gobbledygook.

M. Bunge, *Evaluating Philosophies*, Boston Studies in the Philosophy of Science 295, DOI 10.1007/978-94-007-4408-0_6,

It is also clear that, while useful in telecommunication technologies, Shannon's theory is far too general to help understand anything in physics, biology, or social science. Indeed, this theory is indifferent to the kind of materials that the various parts of a communication system—source, channel, and receiver—are built with. It is not that information systems are immaterial, but that they can be built with different materials, just as a melody can be played on different musical instruments.

Hence the theory in question cannot say anything about the specific *mechanisms* (electromagnetic, chemical, biological, social etc.) involved. Consequently it can *explain* little if anything, since to explain a process is to unveil the underlying mechanism (Bunge 2006).

For example, we say that the nail rusted because its iron atoms combined with oxygen atoms; that the plant grew because its cells divided; that so and so worked well because he was motivated; that the firm prospered because it sold goods on demand at affordable prices; and that that government became unpopular because it was perceived as oppressive and corrupt.

The computational and information-theoretic accounts fail to explain because they do not tell us what happens: they just redescribe in pseudotechnical jargon what could be said in plain language. For example, nothing is learned by translating "She made a throwing motion" into "She computed a throwing motion". There is computation proper only when some algorithm is involved. Hence the increasingly popular slogan, *To be is to compute or to be computed*, is at best false.

Worse, translations into Computerese can be misleading, as when neuroscientists look for algorithms, rather than laws of nature, in the nervous system. Or when they focus on the transmission of neural signals, forgetting that their effect depends critically upon the state of the neural receptor. The confusion in cognitive neuroscience concerning information is such, that a reader in the field (Rugg 1997) contains, back-to-back, one paper (Chap. 2) that claims to use Shannon's concept, and the following chapter, which warns that neurophysiologists use a different one—which it does not define. Nor is there consensus about the code or codes "used" by neurons and neural systems to communicate among themselves.

We now define the concept of a communication system. We stipulate that the quadruple $\sigma = <C, E, S, M>$ represents a *communication system* if

(i) the *composition* C is the set of material components of σ, in particular a coder and a decoder embodying a code;
(ii) the *environment* E is the set of things outside C that act on, or are acted upon by, elements of C;
(iii) the *structure* S is the collection of ties or bonds among members of C, in particular coded signals;
(iv) the *mechanism* M is the collection of processes involving members of C or E, which cause coded signals to be produced, transmitted or detected by members of C.

Given that there are both natural and artificial signals, we must distinguish artificial (or made) communication systems from natural ones. For example, the genome-proteome systems are natural, whereas the linguistic and web communities are artificial.

6.2 Back to Information

Henceforth we shall deal with communication systems of a special kind, namely information systems. These involve animals, or their artificial proxies, engaged in exchanging ideas or their symbols. Examples: two hominids exchanging grunts; a group of babies who exchange information by babbling and gesturing; a scientific meeting; a parliament; and a system composed of robots bonded by coded signals transmitting ideas of their human users.

Let us examine the key concept of a code. A *code* may be defined as a biunique (one-to-one) correspondence between two sets composed of elements of different nature, such as words and sounds, ideas and symbols, or instructions and actions. (The historically evolved languages, usually called "natural", violate the biuniqueness condition: they are cursed, and also blessed, by ambiguity.)

As for *information* in the strict sense, it may be loosely defined as a property of signs or signals that, when received by an animal, elicit in it mental processes, such as having an idea, feeling an emotion, or deciding to do something—or their counterparts in artificial systems. (In the case of artifacts, the effect will be a physical or chemical process rather than a neural one.) Obviously, not every signal carries information. For example, neither a sound of constant amplitude and frequency, nor a perfectly sinusoidal wave, can "tell" us anything.

The concepts of code and information combine into that of *information code*. The following diagram suggests the gist of this concept (Fig. 6.1).

Let us return for a moment to the distinction between natural and artificial codes. Three of the earliest known natural codes are the wound-pain, food-wellbeing, and lightening-thunder correspondences. The most complex of all natural codes is perhaps the genetic code, or gene-protein correspondence. However, no sooner it was discovered, than it turned out that the "translation" of genetic material into proteins obeys a multiplicity of codes—as if there were separate dictionaries for beautiful and for ugly words. And, of course, whereas the sequences ACGT and TGCA carry different genetic "information" (structure), they are assigned the same quantity of information on Shannon's theory.

Artificial codes come in a large variety. The earliest of them is likely to have been the sounds-ideas code involved in speech. Much later came the code that correlated hieroglyphs, ideograms, or written words, to spoken words. And much later came the dictionary that facilitates the deciphering of foreign languages. However, we

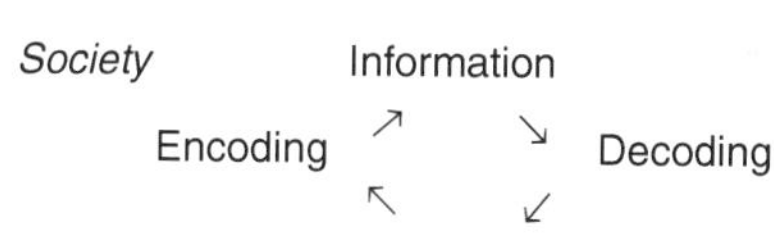

Fig. 6.1 Information codes, such as those involved in speaking and listening, as well as in writing and reading, transform private mental events into information items that can be rendered public and transmitted across space or time

have trespassed the frontiers of information technology, for this discipline deals only with material artifacts of a certain kind, whereas languages are public goods of human groups: no language without speakers.

Finally, we propose defining *information technology*, or *informatics* as the study and design of information systems. In other words, what is common to all the so-called information technologies are the concepts of information code and signal. The concepts of meaning and interpretation belong in semantics, not in informatics.

For this reason, linguistics and library science do not belong in informatics. But of course this will not prevent the information technologist from borrowing concepts from linguistics, semantics, and other disciplines. For example, the designer of automatic translation programs uses the definition of faithful translation as that which preserves meanings. Which illustrates the general maxim, that modern technology builds on basic science (recall Chap. 4).

6.3 Conclusion

Our definition of informatics as the study and design of information systems applies to communications engineering, in particular both hardware and software engineering. An obvious objection to the proposed definition is that it leaves computation out of informatics. This is why the expression "Information and Computation Technology" (ICT) is often preferred to 'Informatics'. However, it should be noted that there are two kinds of computation: on concepts and on signals. The former can be mental, and it may or may not be algorithmic (or rule-directed), whereas computation technology centers on the design of algorithms for carrying out "mechanical" (mindless) computations on symbols and their physical tokens.

Finally, let us address the question of the place of informatics (with or without computation) in the system of knowledge. I suggest that informatics is a hybrid discipline and, more particularly, that it is composed of some mathematical items, such as the general theories of coding and computation; certain technologies, in particular those whose task is to design information systems (computers and programs); and crafts, such as the art of maintenance and repair of information-processing devices. All of these fields share the concepts of code, coded signal, and communication system. By contrast, none of them uses any concept of meaning, which is one reason for leaving linguistics out of informatics.

References

Beynon-Davies, Paul. 2010. The enactment of information: A unified view of information, systems and technology. *European Journal of Information Systems* 19: 389–408.

Bunge, Mario. 2006. *Chasing reality*. Toronto: University of Toronto Press.

Mahner, Martin, and Mario Bunge. 1997. *Foundations of biophilosophy*. Berlin/Heidelberg/New York: Springer.

Rugg, Michael D. (ed.). 1997. *Cognitive neuroscience*. Hove: Psychology Press.

Chapter 7
Wealth and Well-being, Economic Growth and Integral Development

This chapter tackles a bimillennary problem in psychology, ethics, economics, and political philosophy: that of the relation between wealth and well-being. What are they, and should we live for pleasure, or rather seek to live a full and useful life?

This is the ancient dilemma between hedonism, or the cult of pleasure, and eudemonism, or the search for a good life. Economists, almost without exception, have opted for hedonism, but they have not found out what percentage of the goods that ordinary people want are not merchandises. This gap is currently being filled by the psychologists, sociologists, socioeconomists, and other workers in the new "science of happiness" (e.g., Huppert et al. 2005*; Diener and Diener* 2008*; Graham* 2009*). Their main finding, that happiness is not for sale, might surprise the orthodox economists.*

On the social level the former problem, concerning individuals, gets translated into the question of national development: What kind of development should we seek and for whom? In particular, should economic growth be prioritized, or should we promote the simultaneous development of all the sectors of society, including the political and cultural? In either case, should development benefit the chosen few or everybody? And should it enhance the well-being of the individual, and make that of her offspring possible?.

This problem, of course, lies at the intersection of three sciences–psychology, economics, and political science—and two chapters of philosophy—ethics and political philosophy. Consequently, anyone daring to propose original solutions to the problem in question will risk being criticized by experts distributed among these five fields, who are not used to talking to one another.

7.1 Is Happiness for Sale?

Adam Smith's *Wealth of Nations* (1776) was the first modern treatise on economics. Although he regretted poverty, and particularly the fact that in his time there were 500 poor for every rich person, Smith focused on the creation of wealth.

M. Bunge, *Evaluating Philosophies*, Boston Studies in the Philosophy of Science 295,
DOI 10.1007/978-94-007-4408-0_7,

His great book was a paean to the Industrial Revolution that was being born in Manchester and Liverpool, and that a few decades later would deeply affect the rest of the world. This world would end up being divided into two: the industrialized or developed, and the rural or underdeveloped—euphemistically called 'developing'.

Some aristocrats and poets lamented the rapid waning of the "ancient order", which they embellished. This complaint was no mere reactionary nostalgia: surely the starving industrial unemployed sung by Heinrich Heine and mobilized by John Ludd must have been far more unhappy than the overworked feudal serfs of yore. However, the lovers of progress tend to be indifferent to its victims.

Adam Smith's recipe for progress was: seek economic growth through large- scale manufacture and free trade. He took for granted the goal, economic growth, because everyone save the ascetics, hermits, Buddhists, and members of some obscure Protestant sects, seemed to seek happiness through wealth. The U.S. Constitution includes the "search for happiness" among its human rights. Evidently, the framers of the earliest secular and optimistic constitution did not consult the Gospels, Augustine, Martin Luther, Jean Calvin, Thomas Hobbes, Arthur Schopenhauer, Friedrich Nietzsche, or Sigmund Freud.

Smith's recipe for progress has not gone unchallenged. For example, the economist Richard Douthwaite (1999) wrote a book titled *The Growth Illusion: How Economic Growth has Enriched the few, Impoverished the Many, and Endangered the Planet*. And Derek Bok (2010: 206–207), who presided over Harvard for 20 years, sums up his book *The Politics of Happiness* by reporting that "What several investigators have done is to challenge prevailing policies at their core by marshalling evidence to show that the constant preoccupation with growth has *not* helped Americans become any happier over the past 60 years despite the doubling and redoubling of the Gross Domestic Product."

Most economists, however, keep repeating that utility (subjective value) is an increasing function of quantity, and that all increase in wealth enhances the quality of life. They also hold that a sound economy grows at the rate of 3% per year, propelled by population increase, technological advancement, and the acquisitive passion. But they do not tell us where that magic number comes from. Nor are they bothered by the inequality between persons and between nations, which has been increasing over the past four decades. And they could not care less about the exhaustion of mineral resources, desertification, global warming, or squandering of everything, from water to people. There are certainly exceptions in that profession (e.g., Stiglitz et al. 2008), but they are usually confined to pointing out the problem, which is not economic but political.

Let's return to our problem: is it true that welfare increases with wealth? This is where psychologists, social scientists and epidemiologists have started to search for scientific answers, publishing their findings in specialized journals such as *Social Indicators Research* and *Journal of Happiness Studies*, as well as in a number of edited volumes. Here are the main results of those investigations.

1. Most people do not know exactly what makes them happy, or at least satisfied. Many keep saying that money brings happiness because it allows them to buy

things they desire; but as a matter of fact this is not how quality of life increases, since greater wealth usually involves greater stress, less leisure, increased personal debt, and other negatives. Those who regard personal happiness as the top good cannot reach it because the requisite effort involves excessive stress. In other words, "the more people value happiness, the more likely they will feel disappointed" (Mauss et al. 2011). Worse, the pursuit of one's happiness may cause the unhappiness of others—as in the cases of Don Juan and Bernie Madoff.

2. Satisfaction, or declared wellbeing, increases with income. Until recently it was thought that this increase ceases on attaining the plateau corresponding to the state where the individual has met all his basic needs and no longer has any anxieties related to money. But not everyone feels satisfied with what is enough for them: many wish to have much more, so that the plateau hypothesis does not hold for everyone.
3. Nobody ever becomes satisfied in every regard. According to the 2008 Gallup-Healthways poll, Americans assign 82.5 points to satisfaction of basic needs, but only 49.6 to their quality of life. Of the four factors investigated—emotional health, physical health, healthy behaviors, and work environment—the latter got the lowest mark, 48.1. This may be because workers seldom have a say in altering that environment. But cooperativists value highly the chance of participating in the planning of their own activities (see, e.g., Zwerdlin 1980).
4. As recent research shows, the most important contributions to personal happiness are good health, freedom to do what needs to be done, maintaining good intimate and social relations, enjoying a good reputation, and contributing to altruistic causes. In particular, volunteering in religious organizations gives more satisfaction because of its involving friendly relations with equals rather than because of the beliefs involved (Lim and Putnam 2010).
5. When people are asked to assign a value between 0 and 10 to their level of satisfaction, an average of 5.8 results for all 130 countries investigated. This uniformity sounds paradoxical and casts suspicion upon the validity of the questionnaire.
6. Satisfaction decreases as GDP (gross domestic product) increases: this is the so-called "unhappy growth" paradox (Lora and Chaparro 2008). Presumably, this inverse relation is due to the social disturbances, in particular the economic inequalities, caused by rapid economic growth when unaccompanied by social development. When the tide rises all the yachts rise, as President Kennedy once said—but those without yatches risk drowning.
7. Ed Diener and others (Diener and Chan 2010) have found that subjective well-being is healthy and therefore increases longevity. But in turn good health contributes strongly to happiness, so that the two factors constitute a virtuous circle: *Health → Happiness → Health.*
8. The New Economics Foundation, based in London, insists that for social policy design, as well as for political mobilization, it would be convenient to have a single number to proxy for well-being (www.neweconomics.org, 2011). Well, there is a measure for a particular kind of *un*wellness, namely stress: this is cortisol level in blood, a parameter that is easy to measure. Assuming that

wellness corresponds to the lowest cortisol level L, an objective measure of unwellness would be $u=(C-L)/C$, where C is the actual cortisol level reading. Hence its dual, namely $w=1-u=L/C$, could be taken to be as an objective wellness indicator. Maximum well-being, $w=1$, corresponds to $C=L$; since the maximum C is roughly $10L$, the minimum well-being value is about $w=1/10$. Caution: all this is tentative.

To summarize, happiness is not for sale: it is a subjective feeling that depends more on health, knowledge and temperament than on circumstances. Of course, the objective conditions of happiness are socioeconomic and political. However, the rise in GDP, by itself, is not one of these conditions.

7.2 Can Well-Being Be Bought?

The first thing that surprises the philosopher who espies the young "science of happiness" is the frequent confusion of happiness with wellbeing, correlative with the conflation of desire with need. Happiness and satisfaction are subjective, but wellbeing is objective. A person may enjoy wellbeing because she is rested and well fed, has friends, works at what she likes and goes places; but she feels unhappy because she has not achieved her major aspirations: her love is unrequited, she does not enjoy the reputation she believes she deserves, does not reside where she would like to, and so on.

Let's clarify the ideas at play. Whereas wellbeing consists in meeting basic needs, happiness consists in fulfilling a person's highest wishes. Moreover, the difference between the present state and the whole past life has to be taken into account. This is why the questionnaires on the degree of satisfaction ask people both how they feel right now and how they evaluate their past life (Kahneman and Riis 2005). Curiously, most Afghans answer "bad" to the first question, and "good" to the second (Graham 2009).

Habituation, too, must be taken into account, since those who have become used to deprivations are the ones who least complain. For example, Amartya Sen has noted that the inhabitants of the very poor Indian state of Andhra Pradesh complain less than those of Kerala. The latter are just as poor as the former, but they are less unequal and are better educated, so they can imagine a better future for themselves and thus are more ambitious.

The difference between objective need and subjective desire, and the correlative difference between wellbeing and happiness, is obvious to psychologists and moral philosophers, but it escapes many economists. Yet, these should be the first to understand it because, unlike happiness, well-being can, to a large extent, be bought. For example, the sad tycoon who checks at a luxurious hotel to commit suicide, buys comfort, though not happiness. By contrast, the peasant who sings on his way to work on someone else's farm feels happy because he thinks that his wage will support his family—and perhaps also because he likes to grow plants.

7.3 The Problem of Inequality

Nearly one century after Adam Smith, Karl Marx, in *Das Kapital*, another milestone in the history of economic theory, shared Smith's admiration for modern technology and large-scale production, but sharply criticized the distribution of wealth, which he regarded as unfair. The science of his great work is certainly dated, because Marx did not use mathematics, and because the economy he described subsists only in some corners of the planet and in the brains of orthodox economists.

But it is also true that Marx's prophecies about globalization, the concentration of capital, and the international conflicts over natural resources and markets have been fulfilled. By contrast, the idyllic picture of capitalism painted by its apologists, such as Ludwig von Mises, Friedrich Hayek ,and Milton Friedman, has turned out to be so shallow and false, that the celebrated financier and philanthropist George Soros (1998) has called it "market fundamentalism."

Marx's moral criticism of capitalism because of the inequalities it consecratesis are valid too, particularly since about 1960, when these inequalities started to increase. His criticism was shared by his contemporaries, the cooperativist Louis Blanc (1847) and the philosopher and economist John Stuart Mill (1965), as well as by John Maynard Keynes (1936), the top economist of the last century. Since then, the study of the inequalities inherent in capitalism has been one of the foci of social science and political philosophy (see, e.g., Tawney 1962; Sen 1973; Bunge 2009). From his Inequality Project at the University of Texas, Austin, James K. Galbraith, son of the famous economist John K. Galbraith, has kept up his father's fight against the academic and political champions of economic privilege (Galbraith and Berner 2001).

Regrettably, Marx's recipe to correct economic inequality was to intensify political inequality: to replace political democracy with the so-called dictatorship of the proletariat in order to demolish the procapitalist state and nationalize the means of production and exchange. This was Marx's great paradox: he aborted modern communism a quarter of a century after having conceived it.

Marx's prescription was tried in the Soviet Union and its satellites with notorious results: swift modernization, transformation of misery into poverty, drastic reduction of income inequality, and mass education, along with political oppression and the accompanying alienation, plus environmental degradation and cultural stagnation after a short period of bloom.

Contrary to a widespread belief, the Soviet Union did not fail because it practiced the famous Socialist maxim "To each according to their needs, and from each according to their abilities" (Blanc 1847). The Soviet regime failed for statizing the means of production instead of socializing them; for being oppressive and preferring loyaly to merit; and for centralizing power in the political class, which in turn was confined to the Communist Party, and paralyzed the economy and the culture. Because it alienated the majority, the pseudosocialist dictatorship lacked popular support and, at the time of its agony, did not find anyone willing to help it. Those who rule by force cannot expect voluntary help.

A decade after the Russian Revolution, Sweden—and shortly after the Second World War the rest of Western Europe—forged a compromise between unbridled capitalism and suffocating statism, namely the so-called *welfare capitalism*. This is a combination of bounded capitalist economy with bounded socialist state. Contrary to the dark prophecies of the champions of the so-called free market, the welfare state has done very well on all fronts: it has a modern and competitive economy, a vibrant culture, a democratic and peaceful politics, and the highest human development in the world (see UNDP 2010).

How was this high level of civilization attained? The earliest and most advanced welfare state was built at about the time of the Great Depression that started in 1929. It was not improvised in the heat of an electoral campaign, but designed by the Stockholm school of economics, headed by the socio economists Gunnar Myrdal, Bertil Ohlin, and Knut Wicksell, and it was built by public servants with the support of the Socialist Party and the workers and farmers unions. Without this grassroots support and control, the new regime might have been a heavy bureaucratic freak.

Right after the Second World War, the British Labour government built something similar, though on a more modest scale. Shortly thereafter the liberals and Christian Democrats of Western Europe tacitly joined in: they spoke of the "social market", and reinforced social assistance in all sectors. Although this regime is often dubbed 'socialist', it is no such thing, for it leaves the corporations in private hands. It is rather capitalism with welfare, or "capitalism with a human face", in contradistinction to the free market–free until the state bails out the firms judged to be "too big to be allowed to fail."

Nowadays the welfare state subsists in all of Western Europe, though always under threat and with large regional differences. In particular, income inequality, measured by the Gini index, is twice as large in the UK (and in the United States) as in Scandinavia or Japan, and it doubled in Russia after the fall of communism. Another important variable, life expectancy, has risen in France from 68 to 78 years since 1970.

I sum, the failure of statist and dictatorial socialism has not vindicated the so-called free market, and it has not invalidated the social justice ideal. This ideal is still legitimate and attainable, as suggested by the fact that the more egalitarian societies are also the healthiest (Wilkinson and Pickett 2009). This holds at both the individual and the social levels. Indeed, marked social inequality is physiologically pernicious, as suggested by observations and experiments performed in the past two decades. These studies show that social exclusion and unjustified subordination cause stress, which increases the cortisol level, which in turn harms the nervous tissue and weakens the immune system to the point of causing sickness (Davidson 2005; Kemeny 2009).

That causal chain, from the macrosocial level to the microsocial one, is supplemented with its converse: sickness causes labor absenteeism, which decreases productivity, and hence GDP. The management experts who know this favor the participation of employees in the organization of their own tasks, as well as investment in health care.

One of the most sensational and famous studies on the relation between inequality and morbility is the one described in the Whitehall I and II reports. They concern the state of health of the British public servants, all of them permanent, well paid, and with access to the same national health services. The studies show that their morbidity increased, and life expectancy decreased, as their rank in the hierarchy descended. The inference is that the lack of control over work, or restrictions on positive freedom, makes people sick.

The authors of the second Whitehall study (Marmot et al. 1991) recommended paying more attention to the social environment, job design, and the consequences of income inequality. Williams and Collins (1995) found that in the United States, morbility and mortality rates depend strongly on both class and race—in the latter case through the income inequalities derived from racial segregation.

7.4 Sectoral Growth and Integral Development

The builders of the welfare state planned a more equitable distribution of wealth, enhanced economic security, strengthened public health care, supported culture, reinforced education, and invigorated participative democracy. In short, the welfare of the people resulted from *integral or systemic, not sectoral, development.*

The prosperity of nations with welfare state refutes the classical (and neoliberal) formulas

$$Welfare = Wealth \tag{7.1}$$

$$Progress = Economic\ growth, \tag{7.2}$$

which imply that the GDP (gross domestic product) per capita is the most reliable indicator of the quality of life. According to this, the inhabitants of Kuwait, Brunei, Quatar, and the United Arab Emirates would enjoy a well-being equivalent to that of Swedes, Australians, the Dutch, and Canadians. But this conclusion is false, since most of the people in those oil principalities are immigrants who work from dawn to sunset and live in abject poverty, as well as being deprived of rights. In short, statistics show that

$$Wealth\ of\ nations \neq Welfare\ of\ peoples.$$

The 1960s saw the emergence of the social indicators movement (e.g., Sheldon and Moore 1968). This intellectual movement challenged the ruling economism, and in 1974 it acquired its own specialized journal, *Social Indicators Research*. That same year, UNESCO convened a symposium where the standard Eqs. 7.1 and 7.2 given above were discussed, and a vector with four components—biological, economic, cultural and political—was proposed as a development indicator (Bunge 1974, 1981).

The economists did not take notice: most of them remained stuck in the economist dogma. They did not even admit that GDP is a defective measure of national wealth, since it includes polluting activities and expenses in security and military aggression. James Tobin and William Nordhaus suggested long ago that a better measure of national wealth is the net domestic gross product, or NDP, equal to the difference between "goods" and "bads" (Klein 2007; Stiglitz et al. 2008).

Worse yet, the economists invaded the other branches of the social sciences, from anthropology to political science to culture studies. They exported their central dogma, that all humans behave so as to maximize their expected utility, whether by smoking or by abstaining from smoking, by marrying or by remaining single, by voting or by not voting, by committing crimes or by staying honest, and so on (e.g., Becker 1976). Thus, whatever the economists said was irrefutable—a mark of pseudoscience.

On rewarding Gary Becker, Thomas Schelling, and other self-described "economic imperialists", the Bank of Sweden rewarded the dogmatism, social insensitivity and political conservatism of those fantasists (see Bunge 1998).

At the same time it also unwittingly rewarded the philosophers who, like Karl Popper (1967), adopted the "rationality principle" even while oscillating between the theses that it is a mere tautology, that it is false, and that it is true.

In 1989 the U.N. Development Program officially abandoned the socieconomic orthodoxy, and proposed its own human development index, or HDI. This is the average of three indicators: GDP per capita, life expectancy, and education level (see details in UNDP 2006). The publication of the value of this new index for most nations ought to have amazed the partisans of economism, for it showed that the correlation of the HDI with the GDP is weak. For example, it turned out that Cuba, an economically poor nation, occupies 50th place, whereas Saudi Arabia, the richest nation in the world, ranks 76th.

Although the new index (HDI) is much superior to the previous one (GDP), it is not good enough, as it fails to include three variables that many social scientists take to be crucial: equality, democracy, and sustainability. Arguably, equality is good as a means to protect health and secure liberty; democracy allows people to fight for their wellbeing; and no social advancement is worth much unless it is sustainable.

The environmentalist or Green movement has been warning that economic growth at the present rate is unsustainable, since it is consuming non-renewable resources, such as soil, water, and minerals, as well as polluting the environment. Yet, most economists refuse to listen, and keep repeating that it is necessary for all economies to grow at least 3% per year. Necessary for what and for whom? Obviously, so that the managerial class may keep its current incomes. Robert Solow, who earned the 1987 Nobel Prize for his work on economic growth, recently committed the heresy of recommending the deceleration of economic activity. Peter Victor (2008), an economist who favors a stationary state for the economies of the developed nations, agrees. Arguably, only the underdeveloped nations have the right to grow until meeting the basic needs of their peoples. Even so, they ought to compensate for the bad side effects of such development with a more equitable distribution of wealth and a more intensive popular participation in politics.

The preceding criticisms suggest replacing the United Nations' HDI with the following far more comprehensive index, *C*, of *integral development* or *civilization* (Bunge 2009: 395):

$$C = (1/5)(H + K + ES + D + S),$$

where
H = life expectancy at birth,
K = adult literacy and school enrolment,
ES = economic security,
D = democratic development index,
S = sustainable environmental development index.

The first two indices are explained in the 2006 UNDP report. The third is defined thus:

$$ES = \text{GDP} \times \text{Employment rate} \times (1 - \text{Gini index}).$$

The last two are defined in Bunge (2009), with the proviso that *S* is still the subject of work in progress.

Presumably, in the scale of civilization levels deriving from the new index, a handful of nations, in particular the Scandinavian ones, would retain their present ranas; others, such as Costa Rica and Uruguay, would be promoted; and others, such as the UK and the USA, would be demoted because of their low level of economic security; finally others, such as China and Cuba, would descend due to their low level of democratic development.

7.5 Conclusions

When speaking of *development*, one should clarify what one means by it and what it is wanted for. This leads to recollecting the difference between happiness and wellbeing—between meeting basic needs and fulfilling aspirations. In turn, this distinction reminds us that not all goods are for sale; and that some of them—such as security, environmental protection, public health, and education—are indivisible (social), and so important and so expensive, that only an economically powerful state, along with a vigorous and free civil society, can provide and defend them.

Exclusive attention to market goods presupposes a mean conception of human nature (*homo oeconomicus*), as well as a biased view of development as exclusively economic growth in which the collective goods, such as peace and prosperity, are just as important as the bads, such as war and national debt. The economists's conception also leads to impoverishing the lives of the many, as well as to the rapid exhaustion of non-renewable resources, from minerals to poets.

There are three possible reactions to the mortal danger implied by the economists's view of life and development: global economic contraction, stationary economy (zero growth), and integral development. Global contraction is neither socially fair nor politically viable, as it ignores that 80% of the world population still lacks what it takes to meet basic needs—a huge task that requires economic growth in many places. It is immoral and ridiculous to ask the homeless not to seek a roof.

The stationary or zero growth economy preached by Daly and Farley (2010), can only be proposed as one point of a broad political program including great social benefits in health, education and leisure, that would largely compensate for the decrease in income. It is naive to preach zero growth in a political vacuum, as it is to limit a political program to a single point, however important it may be. That's why the Greens will never govern: because, although their defense of the environment is admirable, people have more pressing and more varied needs.

The goal of development ought to be the well-being of the many rather than the disproportionate wealth of a few in a few nations. And what is needed to attain a high degree of civilization is a balanced development: a just social order and an advanced culture, together with an economy and a politics in the public service, without waste or massive unemployment, neither wars nor exploitation of individuals or nations. Nothing of this will be achieved with obsolete ideological slogans. To achieve universal and sustainable well-being we must combine the social sciences and technologies with a civil society and a public governance that ensure the obligations as well as the rights of everyone. For this is what justice is: the balance of rights and duties.

Recall that satisfaction, or the feeling of *personal well-being*, is subjective. Still, it has objective indicators. The simplest indicator of personal well-being may be the fraction of the week employed in agreeable or neutral activities. But maybe the cognitive neuroscientists will eventually come up with a physiological indicator measurable with precision in the clinic. In the meantime the unhappy individual could do much worse than trying Betrand Russell's (1933) wise advice: keep learning, get a useful job, think of others more than yourself, and cultivate the company of people for whom feel affection.

Let's now leap from the individual to society. The *collective welfare* B of a region is not measured by its natural resources, its population, the intensity of its economic activity (GDP), or its Dow Jones index, but can be defined by the following, more inclusive, index: the product of the population P (or fraction of the world population) of the region, and its degree C of civilization (see the discussion above). That is,

$$B = P \times C.$$

This is a social or collective indicator: it does not measure individual satisfaction, or the feeling of personal well-being or happiness. Still, it is obvious that a high level of collective well-being facilitates the attainment of a high level of personal well-being.

Finally, the *wealth* W of a region can be defined as

$$W = \textit{Natural resources} + \textit{Collective welfare} - \textit{Security expenses},$$

where the last item is the fraction of GDP devoted to the armed forces, the organs of internal security, and foreign espionage, and subversion. The natural resources have yet to be satisfactorily quantitated; so far we know only how much it costs to import them. And even this cost is inaccurate, because it does not include the military and political expenses incurred in securing some of them.

Evidently, it won't be armchair politicians such as this author who will bell the big cat in question. Still, as the astrophysicist Enrique Gaviola used to say, one should throw ideas of public interest into the wind, in the hope that some of them may germinate somewhere at some time. Even though ideas do not replace deeds, they always guide or misguide them.

References

Becker, Gary S. 1976. *The economic approach to human behavior*. Chicago: University of Chicago Press.

Blanc, Louis. 1847 [1839]. *L'organisation du travail*, 5th ed. Paris: Société de l'Industrie Fraternelle.

Bok, Derek. 2010. *The politics of happiness*. Princeton: Princeton University Press.

Bunge, Mario. 1974. The methodology of development indicators. UNESCO, Methods and Analysis Division, Dept. of Social Sciences. Rev. version Development indicators. *Social Indicators Research* 9: 369–385, 1981.

Bunge, Mario. 1998. *The social sciences under discussion*. Toronto: University of Toronto Press.

Bunge, Mario. 2009. *Political philosophy*. New Brunswick: Transactions.

Bunge, Mario. 2010. *El socialismo*. Barcelona/Buenos Aires: SinPermiso.

Daly, Herman E. and Joshua, Farley. 2010. *Ecological economics*, 2a.ed. Washington, DC: Island Press.

Davidson, Richard J. 2005. Well-being and affective style-neural substrates and biobehavioural correlates. In Huppert et al., eds., 107–139.

Diener, Ed, and Robert Biswas Diener. 2008. *Happiness*. Malden: Blackwell.

Diener, Ed. and Micaela Y. Chan. 2010. Happy people live longer. SSRN-working paper http://ssrn.com/abstract=1701957

Douthwaite, Richard. 1999. *The growth illusion: How economic growth has enriched the few, impoverished the many, and endangered the planet*, 2a.ed. Gabriola Island: New Society Publishers.

Galbraith, James K., and Maureen Berner. 2001. *Inequality and industrial change*. Cambridge: Cambridge University Press.

Graham, Carol. 2009. *Happiness around the world*. Oxford: Oxford University Press.

Huppert, Felicia. 2005. Positive mental health in individuals and populations. In Huppert, Baylis and Keverne, compils., pp. 307–342.

Huppert, Felicia A., Nick Baylis, and Barry Keverne (eds.). 2005. *The science of well-being*. Oxford: Oxford University Press.

Kahneman, Daniel, and Jason Riis. 2005. Living, and thinking about it: two perspectives on life. In Huppert et al., eds., pp. 285–234.

Kemeny, Margaret. 2009. Psychobiological responses to social threat: Evolution of a psychological model in psychoneuroimmunology. *Brain, Behavior, and Immunity* 23: 1–9.

Keynes, John Maynard. 1973 [1936]. *The general theory of employment, interest and money*. En *collected writings*, vols. 7. London: Macmillan and Cambridge University Press.

Klein, Naomi. 2007. *The shock doctrine: The rise of disaster capitalism*. Toronto: A. A. Knopf Canada.

Lim, Chaeyoon, and Robert D. Putnam. 2010. Religion, social networks, and life satisfaction. *American Sociological Review* 75: 914–933.

Lora, Eduardo and Juan Camilo Chaparro. 2008. The conflictive relationship between satisfaction and income. Washington, DC: Inter-American Development Bank, Research Department, Working paper # 642.

Marmot, Michael G., et al. 1991. Health inequalities among British civil servants: The Whitehall II Study. *The Lancet* 337: 1387–1393.

Mauss, Iris B., Maya Tamir, Craig L. Anderson, and Nicole S. Savino. 2011. Paradoxical effects of valuing happiness. *Emotion* 11: 807–815.

Mill, John Stuart. 1965 [1871]. *Principles of political economy*, 7a ed. En *collected works*, vol. 3. Toronto: University of Toronto Press.

Popper, Karl R. 1967. The rationality principle. In David Miller, compil., *Popper selections* (Princeton: Princeton University Press, 1985), pp. 357–365.

Russell, Bertrand. 1933. *The conquest of happiness*. New York: Garden City Publishing.

Sen, Amartya. 1973. *On economic inequality*. New York: Norton.

Sheldon, E.B., and W.W. Moore. 1968. *Indicators of social change*. New York: Russell Sage.

Soros, George. 1998. *The crisis of global capitalism [Open Society Endangered]*. New York: Public Affairs.

Stiglitz, Joseph, Amartya Sen, and Jean-Paul Fitoussi. 2008. *The measurement of economic performance and social progress revisited: Reflections and overview.* Paris: Commission on the measurement of economic performance and social progress. www.stiglitz-sen-fitoussi.fr/en/index.htm

Tawney, R. 1964 [1938]. *Equality*, ed. rev. New York: Barnes & Noble.

UNDP. 2006. *Human development report 2006*. Geneva: UN.

UNDP. 2010. *Human development report 2010*. Geneva: UN

Victor, Peter A. 2008. *Managing without growth: Slower by design, not disaster*. Northampton: Edward Elgar.

Wilkinson, Richard G. 2009. *The spirit level: Why more equal societies almost always do better?* London/New York: Allen Lane.

Williams, David R., and Chiquita Collins. 1995. Socioeconomic and racial differences in health: Patterns and explanations. *Annual Review of Sociology* 21: 349–386.

Zwerdlin, Daniel. 1980. *Workplace democracy*. New York: Harper & Row.

Chapter 8
Can Standard Economic Theory Account for Crises?

The Business Failure Record yearbook shows that about two-thirds of all new American business firms last less than 5 years. And economic historians know that, ever since the Dutch Tulip Bubble of 1637, the economies of all the capitalist nations have failed spectacularly every so often, sometimes dragging down other nations, as happened notably in 1929,1987, and 2008.

Since its inception around 1870, standard economic theory has not dealt with either business failures or global economic crises. It has not even explained them satisfactorily after the event. And yet nearly all the theoretical economists have earned their degrees writing exercises on that theory—which is as if doctorates in chemistry were awarded for monographs on alchemy.

Standard economic theory has not only failed to account for those crises, but it has also greatly contributed to causing some of them. Indeed, it has inspired the prevalent economic policies in the US and elsewhere, particularly as regards the deregulation of the financial market (Griffith-Jones et al. 2010*). So, the question in the title is worth being investigated.*

8.1 Standard Economics Focuses on Equilibrium

The short answer to our initial question should be obvious: Standard economic theory does not account for economic crises because it does not even study them. Indeed, by definition, an economic crisis is a sharp and protracted rupture of economic equilibrium, the state of the economy in which demand is either much greater or much smaller than supply. Now, standard economic theory either postulates or claims to prove that the free market is at all times in equilibrium or very near it: there would never be serious scarcities or gluts. The practical implication is obvious: governments never need to intervene to repair disequilibria—e.g., by bailing out companies that are "too big to be allowed to fail".

For example, Gerard Debreu (1975), Nobel Prize in economics for 1983, started thus his lecture at the International Congress of Mathematicians held in 1974:

M. Bunge, *Evaluating Philosophies*, Boston Studies in the Philosophy of Science 295,
DOI 10.1007/978-94-007-4408-0_8,

"The observed state of an economy can be viewed as an equilibrium resulting from the interaction of a large number of agents with partially conflicting interests." Note that Debreu held that equilibrium is *observed*, not *assumed.* And he said that when the world had not yet recovered from the crisis caused by OPEC the previous year by a sharp increase in the price of crude oil.

Debreu was proud to belong to the tradition of equilibrium economists started a century earlier by Léon Walras and continued by Vilfredo Pareto and John von Neumann. His classical monograph (Debreu 1959) dealt exclusively with economies in equilibrium. And the Royal Bank of Sweden awarded him the most coveted prize for having proved, jointly with Kenneth Arrow—another Nobel laureate—that all the economies with free competition are always in equilibrium.

In short, the mathematical theory of general equilibrium rules out the possibility of significant disequilibria, in particular crises. It describes an ideal world without pronounced scarcities or gluts, let alone serial bankruptcies and the corresponding unemployment It is like a mechanical theory that proved the impossibility of motion, or like a medical school that denied the possibility of serious sickness, much less death.

True, some economists have acknowledged his limitation For example, Paul Samuelson (1965), another Nobel laureate, devoted one of the 12 chapters of his celebrated textbook to economic dynamics. But the attentive reader will note that this chapter is strictly programmatic: all its formulas are just mathematical exercises without an economic interpretation—despite Samuelson's proclaimed devotion to operationalism.

The many readers of that textbook, mostly students born towards the end of the period of expansion of the American postwar economy, did not notice that the chapter was economically empty, perhaps because they had not suffered the Great Depression, the first crisis that undermined the faith in both the free market and its theory—as well as in capitalism.

The equilibrist economists are majority. They are likely to share Milton Friedman's (1991) claim that the standard theory is in good shape: that it has required only formal innovations since it was born more than a century ago in the brains of Walras, Jevons, and Menger. A critical philosopher of economics would say that theoretical stagnation is a mark of senility, not of good health. I would go further, and suggest that economic equilibrism comes near the mental disequilibrium characterized by denial of reality.

An extreme example of such alienation from reality is the theorem due to Robert Aumann (1964), another Nobel laureate, that the general equilibrium theorem also holds for a continuum of traders. This assertion is just as fantastic as Italo Calvino's story of the nonexistent knight, since every human group is finite and therefore denumerable.

Moreover, if traders constituted a continuum, they would not be *agents* proper, since they would be insignificant. (Every member of a continuum, such as a segment of a line, is surrounded by a dense cloud of infinitely many elements similar to it, so that it would have nothing to exchange with them. In the real world, the power of an individual depends on her having something that others lack.) Is it surprising that in 2005 Aumann shared the economics Nobel Prize with Thomas Schelling, another orthodox economist and believer in the rationality of war and nuclear armament?

However, let us go back to the nucleus of standard economics, the general equilibrium theorem. What are its premises? Leaving mathematical niceties aside, it can be said that the theorem follows from two postulates. The first is that of the rationality and freedom of every economic agent; and the second is the assumption that every market is free, that is, competitive. Let us glimpse into both postulates.

8.2 The Economic Rationality Postulate

According to the first postulate, all the economic agents are "rational", that is, they act so as to maximize their expected utilities. In particular, merchants lower prices when demand goes down, and rise them when it goes up; and consumers behave in the dual manner: they buy when prices decrease, and sell when they increase. The second postulate ignores the monopolies that fix prices regardless of demand; the suppliers that force their clients to buy articles they do not need; and the publicity that transforms desires into apparent necessities.

This simplistic theory ignores not only monopolies and monopsomies, but also state regulations and even credit: all transactions would be in cash and instantaneous. But in fact bankers have been just as important for business and governments as entrepreneurs: remember the Medicis and the Fuggers, the Rothschilds and the Morgans. Moreover, the financial sector, which before World War II held scarcely one-fifth of the total wealth in the developed world, nowadays owns nearly half of it. Consequently a financial tremor can cause a tsunami in the remaining sectors of the economy, as happened in 2008.

And yet, orthodox economic theory ignores the increasing importance of the financial sector—not to mention the cooperatives. Of course there are well-paid professors of finance, but apparently only politicians pay attention to them. The successful financier George Soros (2003) has never used their theories, for he regards financiers as no better than alchemists. In particular, they ignore the inherent instability of the securities market. The heterodox economist Hyman Minsky (1986) explained it thus: Prosperity → Massive speculation → Massive bankruptcies → Credit restriction → Economic contraction. Ironically, this cycle does exemplify the market self-regulation thesis, but it also exhibits the unavoidable companion of the pursuit of riches by the few, namely poverty of the many.

Minsky also proposed that the financial market be regulated by the state to hinder speculation. But neither his colleagues nor the central bankers listened to him. In particular, Alan Greenspan, who ran the American central bank for two decades (1987–2006), facilitaded speculation and therefore the latest financial crisis (2007–2010). This he did by opposing the regulation of the financial sector on the strength of the "rational egoism" of his mentor, the pop philosopher Ayn Rand. However, to his credit, Greenspan eventually admitted the failure of this ideology, and recently he even held that corporate taxes had to be raised to correct the US fiscal imbalance. There is nothing like failure to return to reality.

When the 2008 crisis started, the famous professor Paul Samuelson blamed the labor unions for the crisis: in fighting for high salaries they increased prices, and thus caused the crisis. Apparently he did nor know that the current crisis started as a real estate bubble; nor did he know that labor is only about 10% of the cost of a typical manufactured product, and that only one-tenth of the American workforce is unionized. It would seem that mathematical economists do not read newspapers, to avoid being besmirched by reality.

8.3 The Free Market Postulate

Let us take a look at the second postulate, that of the free market. This assumption was falsified by the emergence of the private monopolies, such as the East India Company, as well as by the earliest incursions of the state in the market, such as the Spanish Crown's monopoly on foreign trade during three centuries

Nowadays the economies of all developed nations are mixed. For example, nearly half the GDP of Germany, France and the United Kingdom is controlled by the public sector. And a big slice of the private sector, particularly the heavy industry, is held by a handful of oligopolies. The Marxist economists and the Cambridge school, formed by the disciples of the great Keynes, insisted on the progressive concentration of capital. But they missed the increasing importance of the public sector, not only in the welfare states but in all nations. At any rate, free markets are easier found in textbook than in the real world.

Equilibrium is parallel. Indeed, by definition the equilibrium point is the intersection of the supply and demand curves. But, as Oskar Morgenstern (1963) pointed out, while there are supply curves, the demand curves exist only in textbooks, since one knows at best a few points of the hypothetical curve. Hence, Adam Smith's celebrated invisible hand moves orthodox economists but not economies.

Still, there exist some sharp disequilibria, those that affect any people. Nor is there any doubt that they are harmful. In fact, demand excess causes inflation, which impoverishes everyone. And supply excess causes deflation (except in the sectors dominated by oligopolies), which benefits temporarily the consumer but harms the small businessmen, which in some countries, such as France and Italy, constitute nearly half the adult population.

Contrary to economic orthodoxy, in some sectors, such as the real estate one, it is considered convenient that the supply be several times the demand, to avoid price inflation. Besides, all the important technological innovations may cause unemployment—the most common disequilibrium in the labor market. Finally, all the great public works projects require floating state bonds, which constitutes a disequilibrium of the public treasure.

In short, equilibria are often beneficial, but not always detectable. Sharp disequilibria are at once easily detectable and painful, but the standard theory does no admit their existence. Hence this theory does not fit reality. This violation of the reality postulate, along with the obstinate defense of obviously false assumptions, suffice to indict equilibrium economics as a paradigm of social pseudoscience. (See additional criticisms in Bunge 1998.)

Worse yet, this theory is the basis of the most influential and harmful ideology of our time, namely neoliberalism. This ideology, the brainchild of Friedrich Hayek and Milton Friedman, is summarized in the formula *The market knows best*. An epistemological corollary of this slogan of right-wing politicians, such as Reagan, Thatcher, and Pinochet, is the anti-intellectualist thesis, first advanced by Hayek and Friedman, that scientists and technologists are not to be trusted. The political consequence is that we must "submit to the impersonal forces of the market," as Hayek (1944: 204) put it in his classic anti-Socialist manifesto. Note the curious mixture of economic orthodoxy and egoism with holistic social philosophy and irrationalist epistemology. (See more in Mirowski 2011.)

8.4 Conclusion

The millions of victims of any of the contemporary economic crises are entitled to ask: What is the worth of the theories that led to making or endorsing the decisions of businessmen and statesmen that took us once more to the brink? This is the severe evaluation of Jean-Philippe Bouchaud (2008), head of research of Capital Fund Management, and a physics professor at the Ecole Polytechnique in Paris: In economics "ideas have solidified into dogmas"; these dogmas have "deified" the market; and "free [unregulated] markets are wild markets" rather than the regular self-correcting systems imagined by economic theorists.

Are there alternatives to the obviously failed standard economic theory? In principle there are two: one is to proceed empirically, and the other is to search for better theories. The first course was taken by Nouriel Roubini and a few others who foresaw the bursting of the American real estate "bubble" in 2008. They used their common sense and noted that the house prices had doubled in a few years—surely an indicator of wild speculation bound to lead too massive foreclosures. But occasional commonsensical forecasting is no substitute for true theory.

Are there alternatives to standard economic theory? Steve Keen (2001) believes that a good theory could result from cobbling together the valid fragments of the existing theories, including the Austrian. But which are such valid fragments? Even assuming that there are some great gems there, potent theories are not crafted from shipwrecks but by restarting *da capo*, on the strength of a handful of deep and realist ideas.

Presumably, one of these founding conjectures is that the economy is not a formless market but a social system, and moreover one strongly coupled to the polity and the culture, as well as to the surrounding nature (see Smelser and Swedberg 1994; Bunge 2004). In turn, this ontological thesis suggests the methodological rule: Construct and test socio-economico-politico-cultural theories, that is, theories that interrelate social, economic, political, and cultural variables and their rates of change.

References

Aumann, Robert J. 1964. Markets with a continuum of traders. *Econometrica* 32: 39–50.

Bouchaud, Jean-Philippe. 2008. Economics needs a scientific revolution. *Nature* 455: 1181.

Bunge, Mario. 1998. *Social science under debate*. Toronto: University of Toronto Press.

Bunge, Mario. 2004. *Emergence and convergence*. Toronto: University of Toronto Press.

Bunge, Mario. 2009. *Political philosophy*. New Brunswick: Transaction.

Debreu, Gerar. 1959. *Theory of value: An axiomatic analysis of economic equilibrium*. New Haven: Yale University Press.

Debreu, Gerard. 1975. Four aspects of the mathematical theory of economic equilibrium. In: *Proceedings of the international congress of mathematicians*, vol. I, 65–77. Ottawa: Canadian Mathematical Congress.

Friedman, Milton. 1991. Old wine in new bottles. *The Economic Journal* 101: 33–40.

Gintis, Herbert, Samuel Bowles, Robert Boyd, and Ernst Fehr (eds.). 2005. *Moral sentiments and material interests: The foundations of cooperation in economic life*. Cambridge, MA: MIT Press.

Griffith-Jones, Stephany, José A. Ocampo and Joseph E. Stiglitz, compils. 2010. *Time for a visible hand*. New York: Oxford University Press

Keen, Steve. 2001. *Debunking economics: The naked emperor of the social sciences*. London/New York: Zed Books.

Minsky, Hyman. 1986. *Stabilizing an unstable economy*. New York: McGraw-Hill.

Mirowski, Philip. 2011. *Science-mart: Privatizing American science*. Cambridge, MA: Harvard University Press.

Morgenstern, Oskar. 1963. *On the accuracy of economic observations*, 2nd ed. Princeton: Princeton University Press.

Samuelson, Paul A. 1965. *Foundations of economic analysis*. Cambridge, MA: Harvard University Press.

Smelser, Neil J., and Richard Swedberg (eds.). 1994. *Handbook of economic sociology*. New York/Princeton: Russell Sage Foundation/Princeton University Press.

Soros, George. 2003. *The alchemy of finance*. New York: Wiley.

Chapter 9
Marxist Philosophy: Promise and Reality

Marxism has been ignored, criticized, travestied, and vilified during more than a century. But no one can honestly deny that Karl Marx was a great economist, as well as a courageous social critic and labor organizer, a brilliant journalist, and the deepest and most eloquent critic of capitalism. Even the pope Benedict XVI admitted that Marx has fascinated and continues to fascinate.

Ironically, no one doubts that Marx and his coworker Friedrich Engels also crafted an original and broad philosophical system, namely dialectical materialism, which in their own estimation "stood Hegel on his head," in that it kept the master's dialectics while rejecting his idealism.

In this chapter it will be argued that there is no such philosophical system but just a collection of imprecise statements that fit into one page; that Marx and Engels, as well as their followers, failed to understand that the worse trait of Hegel's system was not its idealism but its confusions and obscurantism; and that their own philosophical excursions were coarse and at variance with the logic, mathematics, and physics of their time.

These flaws, along with their failure to put philosophical conjectures to the test, explain why contemporary Marxism is fragmented into a number of currents. The same flaws also explain why none of the neo-Marxists has made any original contributions to logic, semantics, ontology, epistemology, methodology, value theory, ethics, or political philosophy. In sum, Marxism has become obsolete and barren scholasticism, unworthy of the innovative and emancipating intentions of its founders.

I suggest that, to correctly evaluate and enrich the legacy of Marx and Engels, one should examine its main ideas in the light of contemporary science and philosophy, as well as in that of the history of socialism, or what passes for such. So far as I know, nobody has attempted this big task, perhaps because Marx's thought has attracted mostly fanatics on both sides of the ideological fence. Here I will confine myself to criticizing what I take to be the main ideas of the philosophy of Marx and Engels, without regard to the uncounted emendations and embellishments added by their followers.

M. Bunge, *Evaluating Philosophies*, Boston Studies in the Philosophy of Science 295, DOI 10.1007/978-94-007-4408-0_9,

9.1 Dialectical Materialism

The philosophical nucleus of Marxism is dialectical materialism, a combination of materialism with Hegel's dialectics. This doctrine does not tell us what matter is, but exalts dialectics as the theory of everything. The so-called laws of dialectics, such as formulated by Engels (1940, 1954) and Lenin (1947, 1981), are false insofar as they are intelligible. Let us take a quick look at them.

The first "law" of dialectics, that of the "transformation of quantity into quality", makes no sense as stated. Indeed, every quantity is either the numerosity of a set or the degree or intensity of a trait, property, or quality, as in "one meter long." What is meaningful and true is the assertion that there are critical points, such as melting, and even changes of species, such as the transmutation of elements.

The second "law", that of the "struggle and unity of opposites," is falsified by the existence of simple things, such as electrons and photons. And few if any complex things or systems, such as organisms and business firms, are decomposable into opposites. For example, normal persons are seldom at odds with themselves, and social groups do not split the moment they emerge. Even in deeply divided societies, collaborative work takes much more time than fight, and the struggle for existence trumps class conflict.

The idea that change is contradictory is an instance of thinking in opposites, typical of primitive and archaic thinking. Thus Zeno of Elea believed that change is contradictory: the flying arrow is there and is not there—which is impossible. Of course the physicists who came shortly after Zeno were not hampered by this difficulty: to them, rest was just zero speed, a particular case of motion, not its opposite. Yet 25 centuries later, Engels (1954), echoing Hegel, repeated Zeno's reasoning, except for the conclusion: Whereas Zeno denied change along with contradiction, the modern dialectical philosophers embraced both. Thus they tacitly repudiated two and a half millennia of science.

In particular, Engels (1940, chapter III) found it necessary to alter classical mechanics so as to fit dialectics. Indeed, he resurrected the Empedocles-Kant's conception of matter as the unity of attraction and repulsion; he confused force with energy; and he thought that a planet was driven by a tangential force, rather than by the Sun-planet gravitational interaction. In short, Engels misunderstood the most important physical theory of his time because he worshipped Hegel and despised Newton. (By contrast, his brilliant essay on "The part played by labor in the transition from ape to man", included in the same volume, was a precursor of the current biosociological view of human evolution.)

On top of being confused and of having as many counterexamples as instances, the second "law" is politically risky, because it places struggle and destruction before cooperation and construction. It is well known that struggle is neither the only nor the best means to attain social goals: it is only the last resort. Cooperation is the least risky alternative, and the one likely to give the more lasting results, because it benefits and attracts the greater number. Confucius does not get along with Hegel.

As for the third dialectical "law," that of the "negation" of "negation", it is incomprehensible as long as we are not told what ontic "negation" and "sublation"

(*Aufhebung*) mean. In any case, I hope that my grandchildren do not believe that they are the double negations of their grandpa.

To sum up, the "laws" of dialectics are not the universal laws of change, valid for matter of all kinds—physical, chemical, biotic, social, and technical. We do know a number of *general* laws, valid for bodies, fields, cells, or small social groups. Whether there are any *universal* laws, that is, patterns of matter of any kind, is an open question. Should there be such laws, they would certainly not be found through Hegel's method, namely navel contemplation behind the back of science.

9.2 Hegel's Disastrous Legacy

Ontology is serious business—no less than the very nucleus of any philosophical system. And a serious ontology cannot be condensed into three sentences, much less if these are cryptic. Obscurity is tolerable at the beginning of an inquiry—e.g., among the pre-Socratics; but it is intolerable after Aristotle. And yet, Hegel won admiration by writing such nonsense as "becoming is the synthesis of being and non-being."

I suggest that any Marxist who realizes that Hegelianism is unrecyclable, must choose between conceding defeat, and attempting to clean up and develop whatever can be salvaged from the shipwreck. The first option is perfectly honorable, in view of the fact that Marx himself gave up philosophy when he undertook to write his masterpiece; and that even Etienne Balibar, a prominent Marxist philosopher, thinks that perhaps Marx was "the greatest antiphilosopher of the modern age".

On the other hand, the philosophical legacy of Marx and Engels may be respected as a failed attempt to combine Hegel's "method" with Ludwig Feuerbach's sketchy materialism, as well as with four radical and promising buds of the radical French Enlightenment: systemic materialism (Holbach), evolutionism (Buffon), scientism (Condorcet), and egalitarianism (Rousseau).

Presumably, any attempt to reconstruct Marxist philosophy will have to start by jettisoning Hegel, whom I regard as the most obscurantist of all the Counter-Enlightenment philosophers. This is because he rejected the entire science of his time, and attempted to pass off obscurity and even nonsense as depth—which makes him the prophet of post-modernism.

Jettisoning Hegel won't be easy for Marxists because Marx, Engels and Lenin were awed by Hegel. Indeed, Marx called him "a mighty thinker". Engels admired Hegel while he despised Newton; he spied the science of his day to confirm his conjecture that Hegel had been its prophet; and he attacked Eugen Dühring, a marginal dilettante, instead of writing an *Anti-Hegel*. Lenin devoted the year 1914 of his Zürich exile to study and annotate Hegel's "big" *Logic*. And Mao Zedong wrote his classic on contradiction the very year that the second Sino-Japanese invasion started.

Admiration for Hegel was understandable in Germany a century earlier: He had built a system that seemed to have ready answers to everything and that, at a time of great changes of all kinds, exalted becoming. Besides, Hegel was neither more obscure nor more reactionary than his rivals, Fichte and Schelling. But it is hard to understand that Hegel found admirers in 1914, and even less in 1937, at a time when modern logic, atomic physics, biology and sociology were flourishing.

Attachment to Hegel was so strong, that his first disciples famously split into the Left and the Right Hegelians. And a century later, while Germany was divided into two, there was a Hegel Society, with its corresponding yearbook, in each of the two halves. It is true that Marx and Engels criticized Hegel's idealism, but they did not repudiate his cult of nonsense and his rejection of all modern science from Newton on. Engels even repeated Hegel's absurd claim that Kepler's particular laws entail Newton's general laws of motion.

This contempt of Marxists for formal logic had three unfortunate consequences. One was the tolerance of absurdities as examples of dialectics. Second: attacks on the first Soviet mathematicians who dared work in modern logic. Third: the thesis that mathematical logic is defective in being static, whereas "dialectical logic" is true for being the theory of change. This thesis is triply absurd. First, because it confuses logic with ontology, a conflation that makes sense to an objective idealist like Hegel, but is inadmissible to a materialist. Second, because the scientific study of change, far from being a priori, has a strong empirical component. Third, because "dialectical logic" never got beyond the programmatic phase.

However, these conceptual flaws of dialectics are as nothing compared with its main political product: the Chinese Cultural Revolution (1966–1976), one of the worst horrors in history. Indeed, the rationale for triggering that movement was this: Strife ("contradiction") is the source of all change; now, Communist China had recently solved the main "contradictions": the class war, the conflict with the Kuomintang, and the fight with Japan; hence Chinese society was threatened by stagnation; whence the need for a new great confrontation—this time young vs. old. This is why the Communist leadership incited teenagers to destroy everything old. This call succeed: An estimated 50 million people died, many more lost 12 years of their lives, and whole institutions, starting with schools, had to be rebuilt 12 long years later.

This concludes our cursory examination of dialectical materialism.

(For more on dialectics see Bunge 1981.) Let us now turn to historical materialism, which is supposed to be the application of the former to social matters, yet is actually far more original and fruitful.

9.3 Historical Materialism

Historical materialism is the social ontology that postulates that everything social is economic "in the last instance"—a rider that guarantees infallibility. It also adopts Heraclitus' postulate, that "strife is the father of all." We shall confine our analysis to these two theses—economism and agonism respectively—as well as to holism and the economy/culture dualism.

Holism (or collectivism) is the thesis that the whole precedes and dominates its parts, and must therefore be studied independently of the latter. Aristotle was an early proponent of holism, and Hegel and his followers refloated it shortly after the American and French revolutions proclaimed the supremacy of the individual. In particular, Hegel deified the state, whereas Marx, who hated the state, saw in social class the source of everything individual.

Both positions are logically untenable because, by definition of the part-whole relation, neither exists without the other. This is why one defines the part-whole relation. The theoretical implications of the social ontologies in question are clear: whereas individualism encourages the bottom-up strategy, holism inspires the top-down one. But neither of these strategies succeeds by itself, because all individual agency occurs in a pre-existing system, and systems can only be understood in terms of individual agency—which is what systemism holds. The political concomitants of the two ontologies are just as obvious: whereas individualism encourages asocial orc even antisocial liberalism, holism inspires totalitarianism. Let us now turn to the second postulate: economism.

Economism is the thesis that the economy is the first motor of society. Marx and Engels defended this thesis, also called 'economic imperialism', which nowadays is championed by Gary Becker (1976) y other influential members of the Chicago School—the Third World's bugbear.

I submit that the economist thesis holds only in some cases. For example, the Industrial Revolution and the concomitant rise of modern capitalism changed radically nearly all the aspects of the societies where it happened. But so did certain political changes, such as the independence of the colonies, as well as certain cultural changes, such as literacy and the adoption of Christianity as the state religion.

True, in due course Marx and Engels qualified the economist thesis: in some private letters they admitted that sometimes politics takes the upper hand, and other times culture does it. But this concession came with a bit of idea/matter dualism. Indeed, in his *Anti-Dühring,* Engels divided society into two: the material (economic) infrastructure, and the ideal superstructure (the rest). This is obviously a transposition of the ancient body/mind dualism, from theology and the idealist philosophy of mind, to the social.

I suggest that a consistent materialist should regard the whole of society as a material object, though of course not a physical one, because its components are material objects—animals, mainly people, and artifacts. Besides, the concept of action among material system is well defined, contrary to the concepts of matter-spirit and its dual.

And instead of regarding the economy-culture interactions as instances of the matter-spirit action and its dual, consistent materialists will view them as interactions among thoroughly material entities: persons and social systems. For instance, they will analyze the phrase "technology-economy flux" into "the work of technologists in business firms or government departments."

Furthermore, realists will claim that politics and culture are just as important as the economy, as shown by such innovations as democratization and mass literacy. In other words, they would suggest that in every society we should distinguish, though not detach, three subsystems on the same footing: the economy, the polity, and the culture.

The dogma that the mode of production determines everything else blinds Marxists to the ability of disinterested research to drive technological progress, which in turn feeds industry and the state. Another consequence of that dogma is that Marxists are still debating where to place science and technology: whether in the infrastructure (as a "productive forces"), as Marx once held, or in the superstructure (along with ideology).

Attachment to those dogmas has prevented most Marxists from realizing the complexity of society. Maybe this partial blindness, along with the belief that historical materialism explains everything social without need for further research, accounts for the nonexistence of a Marxist sociology, culturology, and political science. However, let us move on.

Agonism originated in Heraclitus' thesis that "strife is the father of all." Marxist agonism is the thesis that all human history after the birth of civilization is the history of class struggles. This thesis is highly original and partially true, but it ignores important political and cultural processes that have nothing to do with class struggle, such as the Renaissance, the Reform, the Scientific Revolution, Romanticism, and the wars for national independence.

This criticism is not to deny class exploitation and class struggle, but is intended to warn against exaggerations such as that reported by the Marxist Terry Eagleton (2011:152): subeditors of the British Communist newspaper *Daily Worker* "would be handed reports of road accidents with the instruction "Class-angle that, comrade'."

More importantly, agonism overlooks cooperation, which actually is just as pervasive as competition. Indeed, without cooperation there would be no social systems, hence no intra-systemic conflicts, from family quarrels to management-personnel conflicts and political struggles. There is worse to come.

Yet, for all its flaws, the materialist social ontology, once freed from dialectics, in particular of the dogma that conflict is the mother of all change, has been far more fruitful than its idealist counterpart. Indeed, even a cursory review of mainstream contemporary social science literature shows no serious research inspired in the idealist postulate that ideas move society. It is not that they do not matter but that, as Marx wrote, ideas are effective only insofar as they guide or misguide people in getting what they need to survive.

Yet, for all its flaws, the materialist social ontology has been far more fruitful than its idealist counterpart, in particular the *Verstehen* (understanding, interpretation) school. Indeed, because of its focus on conjectured individual intentions, this school has managed to overlook all the macrosocial processes of our time. However, let us move on.

9.4 Epistemology and the Sociology of Knowledge

The young Marx criticized the empiricism of Bacon, Locke and Hume, and praised Kant's thesis of the creativity of the mind. By contrast Friedrich Engels, his coworker, unwittingly adopted the empiricist epistemology originating in the Aristotelian schoolmen who coined the famous formula *There is nothing in the intellect that was not earlier in the senses.*

According to this thesis, even the most abstract mathematical concepts, such as those of zero, infinity, distributivity, and derivative, would be just distillations of percepts. The socialization of this thesis is the extraordinary idea that all ideas, however abstract, "condense the experience of the masses". If this were so, even the religious and pseudoscientific fantasies would be anchored in experience. And no counterintuitive ideas would ever have emerged. For instance, field physics, atomic physics, genetics, evolutionary biology and the quantum theory would never have been conceived.

In sum, Marxism die not enrich epistemology: it only seasoned the empiricist theory with dialectical confusions. Worse, it did not adopt the best contribution of empiricism, namely the scientistic requirement that hypotheses must be put to the test. Ironically, such apriorism contradicted empiricism. On top of this, it led to the ossification of Marxist philosophy. Indeed, most Marxist philosophers argued in the theological style: "This opinion is false because it contradicts what Marx (or Engels) state in that work."

This is how Lenin (1947) reasoned in his otherwise highly interesting *Materialism and Empiriocriticism*—on top of which he accused eminent scientists of being "lackeys of the bourgeoisie." (However, Lenin was right and alone in exposing the subjectivism of Mach and other eminent scientists. No professional philosopher accomplished this much, because the establishment's philosophy was either Kantian or Hegelian.)

While Marx y Engels did not enrich the philosophy of knowledge, they were the grandparents of that peculiar sociology of science that claims that all ideas, far from being brain processes, are social constructions. Indeed, Marx held not only that "social existence determines consciousness", but also that social class generates ideas, and that the individual acquires them through education and tradition. But he did not reveal how a class, which is brainless, could think, let alone how such an outlandish conjecture might be put to the test. (More in Bunge 1999).

In constructing the sociologistic version of the holism of Aristotle and Hegel, Marx and Engels tacitly rejected the materialist thesis that ideation is a brain process. This may help explain why cognitive neuroscience, which puts into practice the materialist philosophy of mind, was born outside the so-called socialist camp.

It is ironic that a school without sociology should have been sociologistic. The sociologistic conception of cognition produced exaggerations such as Boris Hessen's (1931), who, in his famous paper on Newton's *Principia*, claimed that this work was an attempt to solve the technological problems of his time. It did not matter to him that Newton did not invent any machines, but crafted the first grand scientific theory, that encompassed the special laws of motion proposed by Galileo, Kepler, and Huygens. Nor did it matter that this theory became the basis of modern mechanical engineering.

We should not underestimate the social surrounding of scientists, since it will either stimulate or inhibit them. After all, modern science was born in a handful of advanced European countries, not in Spain or in India. But society, however progressive, does no replace the curious and ingenious brain. Nobody, except for the specialized historian, remembers what happened in England in the year 1687—except that it was the year that Newton's *magnum opus* was published.

Marx's thesis that social groups, not individuals, do the thinking, was reinvented a century later by the social constructivists, who hold that all scientific ideas, and even facts, are social constructions (e.g., Latour and Woolgar 1979). The neoliberal version of this holistic view is this: All research is, or ought to be, motivated by profit; and the expert, in particular the investigator, is nobody: the Market knows best. Consequently the universities—the traditional temples of learning—should serve business (see Mirowski 2011 for a trenchant criticism). Thus, tragicomically, Marx and his followers were unwitting precursors of both post-modernism and neo-liberalism. Moral: Beware of Hegel and holism in general.

9.5 Theory and Praxis, Apriorism and Pragmatism

In the last and most famous of his 11 theses on Feuerbach, engraved in the headstone of his grave, the young Marx stated that, while philosophers had "interpreted" the world in various ways, the point is to change it. This statement is ambiguous: it may be read either as the claim that praxis trumps theory, or as a call to social action without social science.

According to either interpretation, the famous formula is a pragmatist maxim. So much so, that the Marxist philosopher Antonio Labriola called historical materialism *the philosophy of praxis*. Presumably other commentators of Marx's have proposed alternative readings. This is the great advantage of ambiguity: it lends itself to multiple readings, hence to endless scholastic disputation.

At any rate, the gist of Thesis XI is that praxis is superior to theory. If this formula is applied to the philosophy of praxis, it turns out that the latter should be replaced with blind action, that is, action without a theoretical backing. In other words, radical pragmatism is suicidal, for it counsels irrational action.

However, Marx did not put his Thesis XI into practice, since he undertook the Herculean task, pursued by Engels, of investigating the capitalist economy of his time. The same cannot be said of all his disciples: In the course of the last century, the social sciences flourished mainly outside the Marxist box—though, alas, often mixed with pseudoscience. The exceptions were the Russian anthropologists and archaeologist (see Trigger 2006), and the British Marxist historians of the second half of the last century (see Barraclough 1979). But none of those scholars crafted an original economic theory or cultivated sociology or politology.

As a consequence, the Communist leadership, both inside and outside the Soviet bloc, lacked a guide based on solid and up to date social science research: it muddled through using simplistic slogans improvised by undisputed leaders. That was a case of uncritical thinking on a world scale.

In particular, the sudden crumbling of the Soviet bloc took all Marxists by surprise, because they had specialized in criticizing capitalism instead of investigating the traits of the so-called "really existing socialism". That is, the dogmatic adherence to ideas that had been original a century earlier blinded them to what was happening under their noses. What more tragic case of mismatch between theory and reality?

9.6 State and Planning

The Marxist theory of the state boils down to the thesis that its only function it to safeguard the interests of the ruling class. This statement does not hold in general because, since its emergence five millennia ago, the state has had two functions: to keep the existing social regime, and to manage the republic or commonwealth, in particular the territory and the public works. In many cases the state has also undertaken to protect the weak and assist the dispossessed. Suffice it to remember the governments of Hammurabi, Pericles, and Asoka, as well as the welfare states of our days.

Neither Marx nor Engels formulated a strategy for seizing power. In agreement with their historical determinism, they believed that the proletarian revolution would be an automatic consequence of the economic "contradiction" of capitalism: "The mode of production rises in rebellion against the form of exchange", as Engels put it cryptically near the end of his *Socialism: Utopian an Scientific* (1880).

Given this view of history as a titanic battle among traits of the economy, rather than among flesh and bone people, it should not be surprising that the founders of Marxism-Leninism had no plans for the day after the revolution. Planning seemed to them a thing for utopians. (Note the underlying fallacy: "If utopians plan, then planning is utopian.") Consequently, the 1917 Revolution caught them without any plans for the construction of socialism.

Lenin claimed that planning the economy would be "as simple as a children's play". The result of such improvisation was that the Bolsheviks started the construction of socialism only one decade after seizing power, when the First Five Year Plan (1928–1933) was started. This plan was all the more successful since it was carried out at the same time that the capitalist economies foundered as a consequence of the Wall Street crash in 1929.

The Soviet planning of the economy was so successful, that it was copied by many pro-capitalist governments. But, because it was excessively centralized, and therefore authoritarian rather than participative, such planning cuts motivation and grassroots responsibility, and overlooks the regional difficulties, that only the local managers can solve.

For these reasons, central planning lost efficiency as the stages in the modernization process of the economy went on, while politics remained stagnant at a pre-democratic stage. This imbalance was a consequence of economism and the infrastructure/superstructure split.

In short, in "real socialism" there was *statization* but not *socialization* of the means of production, along with the concomitant bureaucratization, alienation, and corruption. To put it into Marxist-sounding terms, "real socialism" was corroded by the contradiction between production, a social process, and the exercise of power, confined to the ruling elite.

In spite of all, the Soviet Five Year Plans transformed a basically rural economy into a great industrial power; the quality of life improved enormously while income inequality dropped to the lowest level in the world; and the USSR defeated Nazism and put the Sputnik in orbit. So, it is false that Soviet communism was a total failure. What is true is that, because it was authoritarian, hence elitist rather than egalitarian, that regime was not Socialist, much less Communist.

9.7 Dictatorship and Disaster

In 1848, with Engels's help, Marx invented modern Communism when writing the brilliant *Communist Manifesto*. But he aborted it in 1875 when criticizing the Gotha program of the German Socialist Party. The earlier document (which passed unnoticed in its own day) proclaimed the desirability and inevitability of "universal emancipation". The second document denigrated the "bourgeois" or "formal" liberties and proposed that, when seizing power, the Socialists should install the "dictatorship of the proletariat."

Marx, Engels and their successors assured that this dictatorship would "wither away" in time, but they did not say by what mechanism: their statement was just one of many Marxist dogmas. What is true, is that all states tend to expand, because the state bureaucracy and the security forces have a vested interest in their own preservation.

It is equally obvious that "authoritarian socialism" is an oxymoron, because socializing a good is sharing it, and there is no participation wherever a small minority, such as a political party, imposes by force its interests, ideas, and plans. It should not have surprised anyone that in 1989 the so-called Communist governments in Europe collapsed without firing a shot, moved by some peaceful street demonstrations: the dictatorships had alienated the masses fed up with unjustified inequalities. Thus, paradoxically, the anti-Communist rebellion was done in the name of equality, not only of liberty. Once again, values were shown to come in bundles, not one by one.

9.8 Conclusion

Marxist philosophy was born obsolete because it resulted from the attempt to combine the progressive ideals of the radical French Enlightenment with Hegel's retrograde philosophy. The neo-Marxist philosophers have confined themselves to expanding and commenting on the classical texts, instead of updating them in the light of the new logic and scientific novelties. And most of the Marxist students of society have devoted more time to commenting on those writings than to investigating social problems. While some of them denounced the inequities inherent in capitalism, none of them denounced those inherent in so-called real socialism.

Given the attachment of Marxist scholars to writings they regarded as infallible and forever topical, it should not be surprising that nearly all the great advances in the natural and social sciences during the last century occurred outside the Marxist box, and that some of them, the most revolutionary, were criticized in the name of Marxism. Thus, when ossified, Marxism became a serious obstacle to the advancement of knowledge. At the same time, while the Marxists in power won sensational victories over poverty, cultural backwardness and military aggression, they did not keep the emancipatory promise of Marx and Engels, for they held all the sectors of society under their iron fist instead of fostering popular participation. It never occurred to them that their formula 'democratic centralism' is an oxymoron.

In the face of the theoretical and practical failure of Marxism, the contemporary socialist intellectuals and politicians have only one way forward: To regard Marxism as an early phase, update the philosophy and the ideals of the radical French Enlightenment, and reinvent socialism as integral democracy guided by the social sciences and technologies (Bunge 2009). In short, not *Back to Marx!*, but *Forward from Marx!*

References

Barraclough, Geoffrey. 1979. *Main trends in history*. New York/London: Holmes & Meier.

Becker, Gary S. 1976. *The economic approach to human behavior*. Chicago: University of Chicago Press.

Bunge, Mario. 1981. *Scientific materialism*. Dordrecht/Boston: Reidel.

Bunge, Mario. 1999. *The sociology-philosophy connection*. New Brunswick: Transaction.

Bunge, Mario. 2009. *Political philosophy*. New Brunswick: Transaction.

Eagleton, Terry. 2011. *Why Marx was right*. New Haven & London: Yale University Press.

Engels, Friedrich. 1940 [1876]. *Dialectics of nature*. New York: International Publishers.

Engels, Friedrich. 1954 [1878]. *Anti-Dühring*. Moscow: Foreign Languages Publishing House.

Engels, Friedrich. 1880. *Socialism: Utopian and scientific*. In Marx and Engels, 1986.

Hessen, Boris. 1931. The social and economic roots of Newton's 'Principia'. In N. Bukharin et al., *Science at the cross roads*. Repr. with Foreword by J. Needham and Introduction by P. G. Werksey. London: Frank Cass, 1971.

Latour, Bruno, and Steve Woolgar. 1979. *Laboratory life*. Thousand Oaks: Sage.

Lenin, Vladimir Ilich. 1947 [1908]. *Materialism and empirio-criticism*. Moscow: Foreign Languages Publishing House.

Lenin, Vladimir Ilich. 1981. *Collected works*, vol. 38: *Philosophical notebooks* [1914–1915]. Moscow: Foreign Languages Publishing House.

Mao, Zedong. 2007 [1937]. *Selections: On practice and contradiction*. New York: Verso.

Marx, Karl. 1852. *The eighteenth brumaire of Louis Bonaparte*. In Marx and Engels, 1986.

Marx, Karl. 1859. *A contribution to the critique of political economy*. In Marx and Engels, 1986.

Marx, Karl. 1875. *Critique of the gotha programme*. In Marx and Engels, 1986.

Marx, Karl, and Frederick Engels. 1986. *Selected works*. New York: International Publishers.

Mirowski, Philip. 2011. *Science-mart: Privaizing American science*. Cambridge, MA: Harvard University Press.

Trigger, Bruce G. 2006. *A history of archaeological thought*, 2nd ed. Cambridge: Cambridge University Press.

Wan. Poe Yo-Ze. 2011. *Reframing the social: Emergentist systemism and social theory*. Burlington: Ashgate.

Chapter 10
Rules of Law: Just and Unjust

The rule of law is, of course, the political regime under which most people observe the laws, whether good or bad. We all benefit by the rule of law, for it gives us security, right? Of course, as long as the state itself is not criminal. If it is, if our government is oppressive or predatory, some of us will try to break the law. For example, we will cheat, bribe, hide, or even go into exile. Hence the rule of law is not an absolute good, right? We shall see.

Political crises, in particular regime changes, not only threaten the integrity of persons, but also test the validity of the political, legal, and moral theories: they show which values those theories safeguard or undermine. In particular, the serious political crises, such as coups, elicit fundamental questions about three kinds of legitimacy: legal, political and moral. Let us examine them.

10.1 Politics, Law, and Morals

The law, just like ethics, is about rights and wrongs, and the latter may be moral, legal, or both. For example, murder is condemned by most moral codes, but it is legal wherever the death penalty holds. And most codes of commerce condone speculations that enrich the wealthy at the expense of the poor. What makes the difference between moral and legal rights and wrongs is politics. To put it in metaphorical terms, the law is sandwiched between morals and politics.

Politics is not only the cradle of positive law, or *lex lata*, but is also the site of *lex ferenda,* or law in the making: the workshop where positive law is put to the test and is remade or unmade. The legal positivists' attempt to detach law from politics and morals serves only the purpose of hiding the fact that no legal system can be pure or neutral either politically or morally. It cannot be so, because every legal system handles rights and duties, is motivated by interests and values that could be different, and is imposed, evaded or challenged by political or moral forces. Witness the histories of the abolitionist, labor, and feminist movements.

M. Bunge, *Evaluating Philosophies*, Boston Studies in the Philosophy of Science 295,
DOI 10.1007/978-94-007-4408-0_10,

We proceed to examining the rule of law (*Rechststaat, Estado de Derecho*) from three different but mutually compatible viewpoints: political, legal, and moral. The offshoot is predictable and therefore trivial: there are multiple rules of law. However, contrary to what relativists and legal positivists and pluralists (or formalists) claim, those regimes are not mutually equivalent: some are objectively better than others. That is, some rules of law make it more feasible than others that people may live in peace and enjoy life. And most of us are willing to trade purse and liberty for the security inherent in the rule of law.

10.2 Legal Legitimacy

I suggest that the problem of legal legitimacy is trivial by comparison with the questions of political and moral legitimacy. Indeed, by definition, *Legal legitimacy = Lawfulness = Observance of the law*. Consequently, a power is legally legitimate if it observes the positive laws, whether good or bad, and sees to it that they be observed. In other words, *Legal legitimacy = Rule of law.*

This requirement is minimal, since it can be satisfied by any government issued from a coup that alters he fundamental law, or constitution, to fit it to its own goals, noble or criminal. Note also that the rule of law benefits nearly everyone, because it allows them to get on with their lives without surprises or alarms as long as they are law-abiding citizens. This explains the political passivity of most people. Only fanatics preach permanent rebellion.

10.3 Political Legitimacy

As we just saw, legal legitimacy suits nearly everyone and it is cheap, because it can be attained by any group who seizes power and alters the rules of the civil game to fit their interests. The problem of political legitimacy is much harder, because a power or a political party may be popular, yet advance the interests of a minority—as is the case with right-wing populism; or, on the contrary, it may be unpopular while defending the interests of the many.

The following definition of the concept of political legitimacy in a democracy accounts for the difference between legitimacy and popularity: A government is *politically legitimate* in a democratic regime if (a) it is supported by the majority of the population; (b) it observes the Universal Declaration of Human Rights (United Nations, 1948); and (c) it obeys the norms of the United Nations Charter (1945) concerning foreign relations.

If a freely elected government violatates condition (b) because it oppresses a minority, or condition (c) for aggressing a foreign power nation, it loses political legitimacy and becomes a rogue nation. Of course, a government may be at once politically legitimate and legally illegitimate, for lacking the power required to force people to observe the law. This was the case of the Lebanese governments during the civil war (1975–1990) that devasted Beirut.

The idea of political legitimacy is clear, but its application may be tricky. For example, it has been known since Condorcet that, in a democracy, a group supported only by 25% of the electorate may form government through cobbling an alliance with another two groups, each of which obtained only 20% of the votes. Thus a minority group may head a government with the support of another two minority groups, provided they renege some of their basic principles, and thus betray their voters. Thus legal legitimacy may be attained at the expense of political legitimacy—and survivorship.

Such anomaly may provoke a coup in the case of a presidentialist regime. In the case of a parliamentary regime, the same situation may lead to either of three results: a weak government unable to take important measures; an unstable and therefore short-lived government; the unraveling of the coalition. The sophisticated voters prefer weak but stable governments, and they wait to the next election to punísh the incompetent or corrupt. By contrast, the poorly educated voters prefer a "strong leadership" ready to take important decisions by themselves. Hence the permanent temptation of dictatorship, and the fragility of democracy in a people lacking in political education.

10.4 Moral Legitimacy and Legitimacy Tout Court

A power may be politically legitimate at the same time that it advances the interests of a small minority or harms organizations devoted to the public good, in which case it is morally objectionable.

What is moral legitimacy? I suggest the following definition. An organization is *morally legitimate* if it (a) favors enhancing the welfare (or quality of life) of the population and (b) observes the Universal Declaration of Human Rights (UN 1948).

According to this definition, an organization may be morally legitimate even while being legally or politically illegitimate. Examples: the governments issued from violent actions that, though performed by minorities, liberated colonial peoples or overturned oppressive governments—such as the constituent assemblies formed by the movements of independence of the Americas.

Finally we can introduce the concept of social legitimacy *simpliciter*:

A power is *legitimate* if it is morally, politically, and legally legitimate.

To put it in metaphorical terms, legitimacy is triangular: it is undone if any of its sides vanishes. But is also renewed every one of these is rebuilt. For example, it is far easier to maintain the rule of law in nations with low income inequalities and without a tradition of violence, such as Costa Rica and Sweden, than in Colombia or the USA.

10.5 Emergencies

The quality of governments, like that of cars, is put to the test in emergencias. I suggest that a government is legally correct if it does not resort to emergency measure to face emergencias. In other words, a government is legally good if it maintains the rule of law, even during emergencias, without resorting to massive repression.

It may be objected that this is impossible when an armed group attacks state installations. Answer: Firstly, a legally correct government reacts by taking circumscribed force measures, not by mobilizing or immobilizing the entire population. Secondly, there are no such groups in a minimally just society, because in them people meet their basic needs, so that nobody revolts so as to be able to eat, nobody owns weapons, and nobody attacks an arsenal.

Think of the Swiss citizens who return home carrying the machine guns o missile launchers they have been using during their periodic military exercises. None of them would go a school, a temple, or a market, to use such weapons against civilians. Those citizens have been educated to live in peace with their fellow humans. They instantiate the maxim *Leges sine moribus vanae*. Or compare the reactions to the Katrina hurricane (New Orleáns, 2005) and to the Sendai tsunami (Japan, 2010). In the former there were lootings, troops, and shots. None of these in the latter case: Today's Japanese are solidary and law abiding.

10.6 If You Wish Order, Prepare for Disorder

Let us deal briefly with social crises that sometimes generate violence: scarcity and high cost of food, massive unemployment, and interruption of essential social services. In most cases people tolerate the consequences of climatic catastrophes, such as prolonged draughts, but they rebel against the artificial scarcities provoked by speculators, or against taxes that do not pay for public services.

A well organized state accumulates reserves of water and food to face natural calamities. This has been done since the beginnings of civilization. And there are no artificial scarcities wherever here is a minimum of social justice.

A particularly interesting case of emergency is the public disturbance caused by massive unemployment combined with food scarcity and large scale kleptocracy. This happened in the Arab countries from the start of the year 2011. In this case we would celebrate the transient failure of public order caused by the political and moral failure of the state.

Let us address briefly the problem of the strike of public employees. This is far more serious than the strike in the private sector, because security, health, education and more are in the hands of the public services. In this case, more clearly than in any others, the maxim of the first International Working Men's Association (1864) applies: *Neither duties without rights, nor rights without duties.*

Because public servants have two masters—the state and the public—it has extraordinary duties and, when discharging them satisfactorily, they are entitled to suitable remunerations and working conditions. Consequently, the public servants' tradeunion should take into account that the public participates unintentionally in all the conflicts it may have with the state, and should consequently attempt to minimize the disturbances that the strike may cause users. A fair and effective way of doing so is to call for intermittent work stoppages.

In short, the slogan of the striking public servant should be: "To work less, or else intermittently, as long as the negotiations continue, but keep supplying the basic services." Besides, taxpayers should be reminded that their servants are just as entitled to live as the private sector workers. And politicians should be told that, if the state does not have funds to pay its servants, some expenditures should be cut, or certain taxes should be raised.

The preceding suggests avoiding total strikes in the public sector, and this for both moral and practical reasons. The former because of the big social responsibility of the public servant, and the latter because the electorate does not forgive the potracted cessation of public services. Remember that Margaret Thatcher rose to power in the wake of the long strike of the garbage collectors.

Finally, how to save the rule of law when sedition threatens? Obviously the answer depends upon the kind of government. Whereas an authoritarian government will intensify repression, a democratic one will seek to eliminate the causes, such as abuses of authority, corruption, food hoarding, and outrages of armed militias.

In short, government and people must prepare to face foreseeable calamities: *Si vis ordo, para turba*. And democracy, which is imperilled by emergency legislation, can be saved by popular participation. When the enemies of the French Revolution of 1789 prepared to invade the young republic, the *Marseillaise* calls "To arms, citizens!".

10.7 The Ultimate Test: The Rise of Nazism

The concepts of rule of law and legitimacy defined above should help clarify one of the most complex and sinister processes in contemporary history: the rise of the National Socialist Party to power (1932–33). The history of Nazism as a political movement is rather well known (see Shirer 1960). However, there still is much to be known. For example, in his admirale study of European fascisms, Michael Mann (2004) overlooked the crucial role of big business and big landowners, the churches, judges and intellectuals in the disintegration of parliamentary democracy, a process that prepared the Nazi seizure of power.

We will confine our attention to recalling the contribution of the courts, jurists and legal philosophers to that process. This contribution has been largely ignored by the historians because it was not made by politicians; by the nazis because it refutes the myth that they earned power through the vote; and by the communists because they did not fit in with their simplistic definition of fascism as "the terrorist dictatorship of finance capital".

The Weimar Republic (1918–1933), the first German democratic regime, fell neither by the vote nor by a popular rebellion. It was destroyed in the course of 6 months by two coups staged by the clique headed by President Hindenburg and Chancellor [Prime Minister] von Papen, at the head of the federal power.

The first coup, on July 20th, 1932, was the presidential decree whereby the federal government took over the reins of power in Prussia, the richest and most heavily populated of the German provinces. Although this decree violated provincial autonomy, it was backed by Article 48 of the national Constitution, concerning emergency powers. This constitutional proviso was the worm in Weimar's apple, as it empowered the federal President to dissolve Parliament and resort to the force of arms. (Ironically, the article in question had been introduced by the Socialdemocratic government presided by Friedrich Ebert [1919–1925], with the approval of the military, to facilitate the repression of the left wing of his own party, along with the communists.)

In 1932, the required emergency was engineered by the federal President and his cabinet, who resolved to unseat the government of Prussia, held by the socialdemocrats. The second coup was the decision of the same President Hindenburg, on January 30th, 1933, by which he appointed Hitler *Reichkanzler* even though he had obtained only 37% of the votes in the last election. To repeat: 63% of the electorate had voted against the dictador in waiting. So much for the legend that Nazism sprang from democracy. What is true, is that the whole process took place within the rule of law—but a law that included the infamous Article 48, a tacit invitation to dictatorship.

Back to the first *Putsch*, which gave rise to the last legal and philosophical debate to take place in the Weimar Republic. The socialdemocratic statesmen dismissed from the Prussian government appealed to the competent court. The latter, invoking Article 48 of the Constitution, ruled in favor of the federal government. Thus the socialists were the victims of their own weakness a decade earlier.

To sum up, the political coup of President Hindenburg and the von Papen cabinet was consecrated by the judges. These, like most university professors and students, sympathized with the Nazis, to the point of authorizing their paramilitary organizations while outlawing the Red Front.

The same coup was also justified by two of the most prominent German legal philosophers of the day: the pro-Nazi Carl Schmitt in an enthusiastic manner, and the liberal Hans Kelsen hesitatingly. Only the third legal scholar, the social-democrat Hemann Keller, dissented. Let us see briefly the point where Kelsen differed from Schmitt (see Dyzenhaus 1997). This question is interesting because it exhibits at once the links between legal philosophy and the remaining branches of philosophy, and the political uses of legal philosophy, even though most of its students overlook this matter.

Schmitt applauded the ruling of the Appeals Court, arguing that Article 48 of Wimar Republic confers absolute powers to the federal President in cases of emergency. Kelsen criticized that article because it violated parliamentary democracy, but at the same time he condoned the ruling of the court because the latter did not invalidate the presidential decree. So, in the end Kelsen exculpated the court (Dyzenhaus op. cit.: 127). Hence, as a matter of fact, in spite of his democratic convictions, Kelsen facilitated Hitler's rise to power just as much as the Nazi Schmitt. This is because they shared their main philosophical ideas, in particular relativism, and refused to acknowledge the political implications of the latter.

10.8 Legal Positivism: Fig Leaf of Authoritarianism

Carl Schmitt and Hans Kelsen shared, expanded and practiced the legal positivism of Hobbes, Bentham, Hegel, and Austin. In particular, both were relativists and preached blind obedience to the ruling legal regime, which amounts to political conformism and legal conservatism.

Legal positivists hold that might makes right, which is historically correct. But they also claim that justice is what the positive law mandates, which is morally and politically wrong. Indeed, it entails that, as Hobbes wrote in his *Leviathan* (1914: 149), "Laws are the rules of Just and Unjust; nothing being reputed Unjust, that is not contrary to some Law." Paul, in *Romans* 13, had said it earlier and in terrorist language. The ethical idea of justice is just as alien to Christianity as to legal positivism.

If law were alien to ethics, it would be mistaken to challenge this or that positive law in the name of extralegal principles, such as "Liberty, equality, fraternity." As Kelsen (1967) repeatedly stated, every legal regime is a rule of law. Consequently, if a group were to seize power by force, and were to proclaim a new legal code, the difference between *de facto* and *de jure* governments would cease overnight. By contrast, the social movements that seek to alter the prevalent legal order, as was the case with the liberal, socialist, trade-unionionist and feminist movements, as well as with the movements of penal reform and of abolition of slavery and serfdom, would be legally unjustifiable.

The Nazi experience should have discredited legal positivism once and for all since, although the Nazi regime observed the rule of law, it violated human rights and entrenched monstrous political slogans, such as Carl Schmitt's, that the *Führer* is the Guardian of Law (see Fuller 1958). Yet, legal positivism is still the ruling legal philosophy, perhaps because it sounds natural to the lawyer and the civil servant, who are expected to uphold the law, not to seek justice.

A similar legal school, though one independent from Kelsen and Schmitt, is legal pluralism. According to this school, all legal systems are equivalent, so that there is no reason to prefer some to others. For example, *sharia* law, which condones "honor" killing, would be worth the same as any system hat condemns it; likewise, legal torture would be just as valid as the prohibition to torture.

Thus legal pluralism underwrites the coexistence of civilization and barbarism, science and superstition, the rule of law and custom, isonomy and privilege. Those of us who believe that on the whole modernity is superior to the previous stages of history, and admire the UN Charter of Human Rights, reject legal pluralism as antihumanist and therefore just as inimical to moral and legal progress as legal positivism. We therefore hold that the law, far from being axiologically and ethically "pure", must safeguard the universal values, in particular justice (see Bruera 1945).

The philosophical root of both pluralism and legal positivism is radical value relativism (Kelsen 1967: 66 ff). According to this doctrine, which is prevalent among the so-called postmoderns, all the value judgments are subjective or tribal, so that "anything goes". But if everything were worth the same, nothing would be worthwhile. Hence axiological relativism is nihilistic and it leads to either inaction or despair.

Besides, relativism is false: normal people have strong preferences about everything that matters to everybody: they prefer well-being to penury, equity to inequity, order to disorder, peace to war, solidarity to egoism, and so on. Moreover, one can give good reasons for such preferences. For example, that they favor survival and living together, which in turn are indisputed values. There are, in sum, objective values, and it is worthwhile to work and fight for them.

The UN Universal Declaration of Human Rights (1948) stipulates the minimum conditions for any regime to qualify as a member of the international community. Because such rights are stipulated to be universal, they constitute a tacit condemnation of value relativism, legal positivism, and legal pluralism.

10.9 Conclusion

Any type of human behavior, good or bad, can be legalized: there are good laws and bad laws. The rule of law is desirable to the extent that the laws in question are objectively good, that is, they it protects universal values.

In other words, legal legitimacy is not enough, not even when accompanied by political legitimacy. Moral legitimacy too is necessary, because the good governance of any social group is not confined to keeping order, but helps realize justice defined as the balance of charges or duties with benefits or rights (Bunge 1989: 187 ff).). *Leges sine iustitia malae, et vanae sine moribus.*

References

Brecht, Arnold. 1959. *Political theory*. Princeton: Princeton University Press.

Bruera, José Juan. 1945. La lógica, el Derecho y la Escuela de Viena. *Minerva* 2: 170–177.

Bunge, Mario. 1989. *Treatise on basic philosophy*, vol. 8: Ethics. Dordrecht, Boston & Lancaster: D. Reidel.

Dyzenhaus, David. 1997. Legal theory in the collapse of Weimar: Contemporary lessons? *American Political Science Review* 91: 121–134.

Fuller, Lon L. 1958. Positivism and fidelity to law: A reply to Professor Hart. *Harvard Law Review* 71: 630–672.

Hobbes, Thomas. 1914 [1651]. *Leviathan*. London: J. M. Dent & Sons.

Kelsen, Hans. 1967 [1934]. *Pure theory of law*. Berkeley: University of California Press.

Mann, Michael. 2004. *Fascists*. Cambridge: Cambridge University Press.

Shirer, W.R. 1960. *The rise and fall of the third Reich*. New York: Simon & Schuster.

Part III
Philosophical Gaps

Chapter 11
Subjective Probabilities: Admissible in Science?

Bayesianism is the opinion that probabilities are a matter of opinion. This is because they would only measure the strength of our beliefs (De Finetti 1972; *Jeffreys* 1975; *Keynes* 1957; *Savage* 1954*). That view is generally known as Bayesianism because of its heavy reliance on a certain interpretation of Bayes' theorem, a piece of pure mathematics that actually refers neither to the knowing subject nor to the real world.*

More precisely, according to Bayesians, all and only propositions (or statements) qualify as more or less probable—although they do not bother to explain what is meant by "this proposition is probable", or by "this proposition is more probable than that one."

Bayesians admit that your assessment of the probability of a given proposition is likely to differ from mine. But, because probability, like beauty, would be in the eye of the beholder, there is no way to tell which of us is right. Indeed, as the statistician who started the contemporary phase of this current put it, Bayesians "maintain that a probability, being but a measure of someone's belief, is not susceptible of being proved or disproved by the facts" (De Finetti 1962*: 360). Anything goes!*

My goal in this chapter is to examine Bayesianism to find out whether it is scientific, and thus deserves the attention of scientists, engineers, legal experts, medical doctors, and other specialists. To perform this task I will draw heavily on some earlier work of mine (Bunge 1951, 1955, 1976, 1981, 1988, 2006, 2008, 2010*).*

11.1 Beware Ordinary Language

Bayesianism is popular outside science and technology, largely because of the confusion of the word 'probability' in its mathematical sense with its ordinary-language homologue, which covers such different concepts as those of possibility, frequency, likelihood, and plausibility or verisimilitude.

More than two centuries ago, Bayesianism penetrated the law, where it is sometimes referred to sarcastically as "trial by numbers". Thus, according to one juror in almost

M. Bunge, *Evaluating Philosophies*, Boston Studies in the Philosophy of Science 295,
DOI 10.1007/978-94-007-4408-0_11,

any trial, the defendant is "probably" innocent, whereas another juror swears that the same individual is "probably" guilty. Moreover, while a judge may state confidently that a given criminal is "probably" a recidivist, he may decide that another individual will "probably" desist from a career in crime. Not surprisingly, in Texas the law requires jurors to consider "whether there is a probability that the defendant would commit criminal acts of violence that would constitute a threat to society." Regrettably, the jurors are not instructed on how to evaluate such a "probability": presumably, they are expected to use cowboy's horse commonsense to decide on a person's life.

What to do in the face of such uncertainties? A scientist would presumably recommend reviewing all the evidence of the case, summon further experts, or perhaps checking the circumstances of some jurors to make sure that they have been neither bribed nor intimidated. Not so a learned jurist of the Bayesian persuasion: He will likely (not "probably"!) suggest applying Condorcet's famous Jury Theorem: Increase the number of jurors regardless of their qualifications and the quality of the extant evidence.

But of course the jurist sold on Bayesianism is unlikely to exhibit any empirical evidence for or against the truth of the said theorem, if only because the very concept of truth does not occur in Bayesianism, which is radically subjectivist, hence relativist. So, "our learned friend" practices the law based on no evidence whatsoever. The nose has replaced the frontal cortex, and consequently the kangaroo court has replaced the legitimate court of law.

The conspicuous absence of the concepts of objective truth and evidence should suffice to alert the skeptical jurist to the dangers of Bayesianism. Indeed, the very first step in any legitimate judicial process is to try and find out the truth of the matter: To establish whether or not the facts in question have occurred, and whether or not the relevant statements fit the said facts, that is, are true at least to a first approximation. In the law, just as in science and in technology, we seek solid *episteme* (science) rather than mere *doxa* (opinion).

Moreover, in the field of human action there are certainly plenty of accidents, but there is no irreducible chance or randomness of the kind we meet in quantum mechanics, quantum chemistry, molecular biology, or genetics. In particular, the concept of chance or randomness has no place in criminology. People either steal or they don't, murder or they don't, and so on: There is nothing random about crime, because every criminal action is the last link in one or more causal chains. Now, it so happens that in the sciences only states and changes of state (events) are assigned probabilities. Moreover, often such probabilities can be calculated unambiguously and measured accurately rather than being assigned arbitrarily.

For example, physicists calculate and measure the probability that a given incoming particle will be deflected by a target atomic nucleus within a given solid angle during one time unit. True, probability is also used in the biosocial sciences, such as epidemiology, and in the social sciences, such as sociology. Thus, for instance, one can legitimately speak of contagion probability, and of the probability of someone climbing up or down a social hierarchy on reasons other than merit. But in these cases, just as in natural science, the probabilities in question are objective, not subjective, and they are estimated by observed relative frequencies. That is, the frequencies of certain types can be used as probability indicators (see Chap. 13).

In sum, the interpretation of probability employed in science and technology is not subjective or Bayesian, but objective or realist. It is regarded as a measure of objective possibility, objective disorder, or both—as when extracting a card from a repeatedly shuffled deck of playing cards.

Nobody in science, except for the psychologists and sociologists who study belief, is interested in mere opinions about facts; and nobody is bold or naïve enough to assign probabilities to either scientific data or scientific hypotheses, the way Bayesians do. To be sure, one often indulges in subjective estimates (or eye-balling) of objective probabilities, just as one may perform intuitive estimates of any other magnitude. However, such estimates are extra-theoretical, and never intended to replace well-founded calculations or accurate measurements.

11.2 Monty Hall Brain Teasers

There is a whole spate of Monty Hall problems, also called Marilyn Vos Savant brain-teasers. All of them allegedly involve probabilities. Here is the simplest and best known of them. There are three identical doors in a room. Behind one of the doors there is a prize for the individual who chooses to open that door without trying any of the other two. The problem is: What is the probability that the prize lies behind any given door? Most people will answer unhesitatingly: 1/3.

People who respond in this manner do not realize that there are *two problems* involved here: guessing the probability of the *presence* of the prize behind a given door, and guessing the probability of *opening the right door*. Whereas the second problem is legitimate, the first one is a pseudoproblem, because it presupposes that the prize has been placed at random, whereas actually the choice of door was deliberate. And, even if the door was chosen at random, chance evaporated once the prize was deposited.

Only the second problem, that of guessing the winning door, is legitimate, because the person invited to open a door chooses it at random, rather than in the light of any piece of privileged information. Here randomness lies in the choice, which is a brain process, not in the room. But even in this case it may be argued that actually what is involved is plain uncertainty. To dignify this mental state with the name 'probability' is illegitimate, because it does not belong in any stochastic model in cognitive psychology.

The three-prisoners problem is more famous and more dramatic, but conceptually the same. Three men, Luke, Mark, and Matthew, are locked up in separate cells. They are waiting for their judge to choose two of them for execution. They reason that each of them has the probability 1/3 of saving his neck. Mark, the more crafty of the three, asks the jailer for the name of one of the three prisoners who is sure to be executed. The jailer accepts, and informs him that Luke is one of the two condemned men. Mark feels relieved, because he believes his chance has risen from 1/3 to 1/2.

Is Mark right? No, because the judge has already chosen who are to be put to death, and there is no indication that he did so at random. In this case, then, there is uncertainty in the minds of two men. Moreover, it is true that Mark's uncertainty has decreased a bit, even though we are unable to ascertain by how much. But randomness is neither in the objective situation nor in the minds of any of the characters of the tragedy. Hence there is no warrant to invoke probability in this case either.

11.3 Learned Ignorance

For all its popularity among philosophers, and its throng of faithful among statisticians, Bayesianism is a minority view in the scientific community. The reason scientists have to avoid Bayesianism is that, because it is subjectivist, it invites arbitrary probability assignments to anything—hardly a scientific procedure.

Bayesianism originates in an ontological dogma and two major confusions. The dogma in question is classical determinism or causalism, so brilliantly described by Laplace two centuries ago, and certainly justifiable in his time. This is the thesis that everything happens according to laws that, such as Newton's, have a broad causal domain (see Bunge 1979). If causalism were true, chance would indeed be but a name for our ignorance of causes, so that an omniscient being would be able to dispense with the concept of chance.

However, the basic laws of quantum theory and population genetics are probabilistic, and they do not derive from causal laws. Rather on the contrary, many a macro-law is a law of averages, and it can thus be deduced from probabilistic micro-laws. The second law of classical thermodynamics is a case in point; another is the law of the exponential decay of a sample of radioactive material. So much for the error at the root of the subjectivist interpretation of probability. Let us now turn to the accompanying confusions.

A first confusion is that between propositions (or statements) and the facts they refer to. Suppose, for example, that V designates a random variable, such as the number of points scored in a die throw. Further, call $P(V=v)$ the probability that, on a given occasion, the variable V takes on the particular value v, such as the ace. The proposition "$P(V=v)=1/6$" involves the proposition "$V=v$", but it should not be read as the probability of this proposition, since such an expression makes no clear sense. The gambler knows that the proposition "$P(V=\text{ace})=1/6$" states the probability of the *fact* of getting an ace when throwing a well-shaken dice cup. He is interested in the outcome of a real process characterized by objective disorder—the one resulting from repeated vigorous shaking.

A second major source of Bayesianism is the confusion of objective chance with subjective uncertainty. This is a conflation between an ontological category and a psychological (and epistemological) one. To be sure, this confusion is rather natural, because objective indeterminacy implies subjective uncertainty—though not conversely. For example, while shaking vigorously a dice-cup, every one of the six sides of a die acquires the same chance of coming up when the die is cast. However, once the die is cast, determinacy has replaced indeterminacy, whereas subjective uncertainty remains as long as we do not look at the die. Bayesians have no right to say that the probability that they will *see* an ace is 1/6, because the random process that culminated in this fact is over: *alea jacta est*. If an ace is what came up, the gambler is allowed to look, and his eyesight is normal, he *will* see an ace regardless of his expectations.

Moreover, the gambler's mental process is quite different from the random physical process that she triggers when rolling dice. So much so, that the gambler who ignores the laws of chance is bound to form irrational expectations, such as the popular gambler's fallacy. ("The next throw must be an ace, since no ace occurred in the last five throws".) That is, our expectations may not mirror objective chance. If they did, neither casinos nor lotteries would be profitable. The only way to defeat chance is by cheating. So much so, that rational people abstain from playing roulette, or any other game of chance, if they suspect that the gadgets involved have been tampered with so as to violate probability laws. In short, *probability if and only if objective chance.*

11.4 Probability as Credence

Let us now examine the Bayesian claim that probabilities are credences, or belief intensities. This is an empirical question, hence it cannot be settled a priori—the way subjectivists do. Let us therefore ask cognitive psychologists whether people actually think in accordance with the said calculus when reasoning about uncertain matters. The many experiments of Daniel Kahneman and his students have conclusively shown that our subjective judgments of likelihood and plausibility (or verisimilitude) are very often incorrect, and do not meet the axioms of the probability calculus (Kahneman et al. 1982).

To begin with, when considering a branching process, such as a phylogeny or a decision tree, there are rarely enough data to include all of the possible forks—a condition for ensuring that the sum of the probabilities over all the branches equals unity. Second, we tend to exaggerate the probabilities of certain extremely unlikely events, such as that of winning the lottery. Third, as Venn (1962) noted more than a century ago, fear, greed, wishful thinking, superstition, strong emotion, and association with pleasurable or painful experiences, are among the factors that distort our judgments of the objective likelihood and actual frequency of an event. For example, one can "prove" the occurrence of any paranormal phenomenon by assigning a high prior probability to belief in it.

Finally, consider the following counterexample to the hypothesis that beliefs satisfy the probability laws. One of these laws is "$P(A\ \&\ B) \leq P(A),\ P(B)$". Set A = "Liberty is good", and B = "Equality is good". Libertarians swear by A, egalitarians by B, and neither by both. In my opinion, neither liberty nor equality is by itself a social good, or even viable, because liberty is impossible among the unequal, and forced equality muffles liberty. But it is arguable that the combination of liberty with equality is both viable and desirable (Bunge 2009). So, if we had a plausible theory of beliefs, it would include the dual of the probabilistic inequality, namely, $C(A \& B) \geq C(A), C(B)$, where C stands for credence or strength of belief.

In short, *Perceived likelihood* ≠ *Objective likelihood*. In other words, *Probability* ≠ *Degree of rational belief*.

11.5 Prior Probabilities Are Inscrutable

A condition for the legitimate application of Bayes' theorem is that three out of the four probabilities occurring in the theorem be known. Hence, when the prior probabilities $P(A)$ and $P(B)$ are unknown, as is the case when A = hypothesis h, and B = evidence e, writing $P(h|e)$ and $P(e|h)$ in terms of them, amounts to scribbling squiggles. And yet this is how Bayes' theorem is used in both Bayesian statistics and inductive logic.

For instance, when estimating the probability of an event, or the plausibility of a conjeture, the Bayesian consults a panel of experts. That is, he seeks "a consensus view of informed opinion", just the way one proceeds in everyday life with regard to everyday matters—with the difference that the Bayesian assigns numbers to strengths of belief (see, e.g., Press 1989).

True, the self-styled objectivist Bayesians equate $Pr(h|e)$ to the corresponding frequency—for example, that a positive clinical test is evidence for a certain sickness; but they make up the other "probabilities", in particular $P(h)$. Besides, in equating certain probabilities with frequencies, they violate the Bayesian dogma that probabilities are credences.

Unknown priors, that must be stipulated arbitrarily, do not worry the Bayesian anymore than God's inscrutable designs worries the theologian. Thus Lindley (1976), one of the leaders of the Bayesian school, holds that this difficulty has been "grossly exaggerated". And he adds: "I am often asked if the [Bayesian] method gives the *right* answer: or, more particularly, how do you know if you have got the *right* prior [probability]. My reply is that I don't know what is meant by 'right' in this context. The Bayesian theory is about *coherence*, not about right or wrong" (op. cit: 359). Thus the Bayesian, along with the logical imperialist who only cares about the cogency of arguments, fits in with the reasoning madman.

11.6 Bayesianism Can Be Disastrous

Unsurprisingly, Bayesianism can have catastrophic practical consequences. Let us recall a famous example, that concerns risk assessment. In 1986, when the space shuttle Challenger exploded, the inquiry committee found that the managers of the project had estimated that the failure "probability" was one-thousandth that foreseen by the engineers. But the Committee did not bother to ask how those numbers had been generated: it did not question the idea that, since there were no precedents of failure of space missions, the experts were free to assign numbers to their gut feelings.

Only one member of the Committee, the famous physicist Richard Feynman, dared state that the overly optimistic risk assessment of the managers was due to their concern for public relations. In any case, both numbers in question, the engineers' 1/100, and the managers' 1/100,000, were "subjective probabilities": they represented the individuals' degrees of belief rather than the reliability of the artifact. And yet, it should be obvious that juggling with probabilities is inappropriate in the case of the failure of components of a machine, because this is a causal chain, not a random one—such as a Markov chain, every link of which has an objective probability that depends only upon the preceding link. As it is well known, Feynman himself found one of the faulty links of the Challenger.

In general, disasters—from computer crash to early death, from bankruptcy to biospecies extinction, from hurricane to tsunami, from epidemic to war, and from bank robbery to murder—are nonrandom events, hence unaccountable in probabilistic terms. True, the extinction of species and the decline of ecosystems have often been modeled assuming that these processes are random. But they are not. Indeed, the dominant ecological variables are nonrandom: Think of climate, species rarity, body size, and the presence or absence of carnivores, keystone species (like starfish), as well as of aggressive invaders (such as elephant grass). Hence those stochastic models are unrealistic, and therefore they are misleading tools for designing environmental protection policies. The moral is that probability without randomness can be hazardous to the environment. (See examples of the health risks of Bayesianism in Bunge 2006.)

11.7 Trial by Numbers

Phrases such as 'probability of guilt', 'probability of recidivism', 'balance of probabilities', and 'probability of a just verdict', are rather common in modern legal jargon. This may be due in part to the healthy influence of the skepticism that accompanies British (particularly Scottish) empiricism. I submit that the phrases in question are legitimate as long as the word 'probability' is taken in its ordinary-language sense, but that they invite trouble when it is interpreted as designating the concept elucidated by the mathematical theory of probability. (The same holds for the French and German equivalents of 'probable'.) I suggest that, to avoid trouble, one should speak of *likelihood* in the case of facts, and of *plausibility* in that of hypotheses. Let us take a closer look at this question, which is not merely a terminological one.

Consider the sentence 'He probably committed the crime'. I claim that the adverb should not be taken literally, in the sense of the probability calculus, because crimes are either committed or not committed: they are outcomes of deliberate actions, not of random processes. (This holds even if the criminal lets the dice decide for him.) Hence, the sentence in question should be replaced with this one: 'It is likely that he committed the crime'. And of course no attempt should be made to quantitate this likelihood. The most we can do is to add the statistical information that crimes of the given type occur with such and such frequency among persons of the same characteristics.

Much the same holds of course for the sentences of the form 'Given (or in the light of) that evidence, it is probable that the defendant committed the crime'. Just say that the evidence in question supports the guess (or hypothesis) that the individual in question is guilty. And do not attempt to assign a number to the weight of the evidence, because the scales of justice are not precise. To write formulas such as '$P(h|e) = 0.75$' is at best a waste of time.

What can sometimes be done is to measure the relative frequency of actions of a certain type. For example, it is known that the frequency of criminal offences is age-dependent. This distribution increases from childhood on, peaks at about 17 years, and then declines down to a plateau. The causes of juvenile delinquency are multiple, among them poverty, the greater freedom and more intense urges of adolescents, jointly with the comparatively slow maturation of the prefrontal cortex (see, e.g., Hawkins 1996; Lahey et al. 2003; Maguire et al. 1994; Robinson 2004; Wikström and Sampson 2006).

Caution: Such data cannot be used to construct the prior probability that an adolescent will commit a crime, because crime is not random. Likewise, the fact that nearly half of the world population is Asian is no ground for expecting that the probability that an Icelandic couple will give birth to an Asian child is one-half. Statistics give us collective regularities, not individual dispositions.

11.8 Concluding Remarks

In conclusion, Bayesian statistics and inductive logic are triply wrong: because they assign probabilities to statements; because they conceive of probabilities as subjective; and because they invoke probabilities in the absence of randomness.

Navel contemplation does not help us explore the real world. Nor does invoking arbitrary numbers: this it is just a disguise of ignorance. Subjective experience is a subject for scientific (psychological) investigation, not a surrogate for it. And subjectivism, whether Berkeley's, Kant's, Fichte's, Husserl's, or Bayesian, is a mark of either anti-science or pseudoscience.

Learned ignorance is still ignorance. In particular, medical diagnosis with the help of made-up numbers is not safer than without them; and trial by numbers is no more fair than trial by water or by combat. Thou shalt not gamble with life, justice, or truth.

References

Bunge, Mario. 1951. What is chance? *Science & Society* 15: 209–231.

Bunge, Mario. 1955. Strife about complementarity. *British Journal for the Philosophy of Science* 6: 1–12: 6: 141–154.

Bunge, Mario. 1976. Possibility and probability. In *Foundations of probability theory, statistical inference, and statistical theories of science*, vol. 3, ed. W.L. Harper and C.A. Hooker, 17–34. Boston: D. Reidel.

Bunge, Mario. 1979 [1950]. *Causality: The place of the causal principle in modern science*. New York: Dover.

Bunge, Mario. 1981. Four concepts of probability. *Applied Mathematical Modelling* 5: 306–312.

Bunge, Mario. 1988. Two faces and three masks of probability. In *Probability in the sciences*, ed. E. Agazzi, 27–50. Dordrecht: Reidel.

Bunge, Mario. 2003. *Emergence and convergence: qualitative novelty and the unity of science*. Toronto: University of Toronto Press.

Bunge, Mario. 2006. *Chasing reality: Strife over realism*. Toronto: University of Toronto Press.

Bunge, Mario. 2008. Bayesianism: Science or pseudoscience? *International Review of Victimology* 15: 169–182.

Bunge, Mario. 2009. *Political philosophy*. New Brunswick: Transaction.

Bunge, Mario. 2010. *Matter and mind*, Boston studies in the philosophy of science, vol. 287. Dordrecht/Heidelberg/London/New York: Springer.

Carnap, Rudolf. 1950. *Logical foundations of probability*. Chicago: University of Chicago Press.

De Finetti, Bruno. 1962. Does it make sense to speak of 'good probability appraisers'? In *The scientist speculates: an anthology of partly-baked ideas*, ed. I.J. Good. London: Heinemann.

De Finetti, Bruno. 1972. *Probability, induction, and statistics*. New York: Wiley.

Eddy, Charles. 1982. Probabilistic reasoning in clinical medicine: Problems and opportunities. In *Judgment under uncertainty: Heuristics and biases*, ed. D. Kahneman, P. Slovic, and A. Tversky, 249–267. New York: Cambridge University Press.

Feller, William. 1968. *An introduction to probability theory and its applications*, vol. 1, 3rd ed. New York: Wiley.

Fréchet, Maurice. 1955 [1946]. Les definitions courantes de la probabilité. In *Les mathématiques et le concret*, pp. 157–204. Paris: Presses Universitaires de France.

Hawkins, J. David (ed.). 1996. *Delinquency and crime: Current theories*. Cambridge, UK: Cambridge University Press.

Humphreys, Paul. 1985. Why propensities cannot be probabilities. *Philosophical Review* 94: 557–570.

Jeffreys, Harold. 1975 [1939]. *Scientific inference*, 3d ed. Cambridge, UK: Cambridge University Press.

Kahneman, Daniel, Paul Slovic, and Amos Tversky (eds.). 1982. *Judgment under uncertainty: Heuristic and biases*. Cambridge, UK: Cambridge University Press.

Keynes, John Maynard. 1957 [1921]. *A treatise on probability*. London: Macmillan.

Lahey, Benjamin B., Terrie E. Moffitt, and Avshalom Caspi (eds.). 2003. *Causes of conduct disorder and juvenile delinquency*, 118–148. New York: The Guilford Press.

Lindley, Dennis V. 1976. Bayesian statistics. In *Foundations of probability theory, statistical inference, and statistical theories of science*, vol. 2, ed. W.L. Harper and C.A. Hooker, 353–362. Dordrecht/Boston: D. Reidel.

Maguire, Mike, Rod Morgan, and Robert Reiner (eds.). 1994. *The oxford handbook of criminology*. Oxford: Clarendon.

Press, S. James. 1989. *Bayesian statistics: Principles, models, and applications*. New York: Wiley.

Robinson, Matthew B. 2004. *Why crime? An integrated systems theory of antisocial behavior*. Upper Saddle River: Pearson/Prentice Hall.

Savage, L. James. 1954. *The foundations of statistics*. New York: Wiley.

Venn, John. 1962 [1866]. *The logic of chance*. London: Chelsea Publications Co.

Wikström, Per-Olov, and Robert J. Sampson (eds.). 2006. *Crime and Its explanation: Contexts, mechanisms and development*. Cambridge, UK: Cambridge University Press.

Chapter 12
Does Inductive Logic Work?

Induction is the jumping from the particular to the general—for instance, from sample to population. And inductivism is the philosophical doctrine according to which all the scientific hypotheses are obtained by induction from empirical data—as even the great Bertrand Russell believed. According to the empiricist tradition, the sciences of facts, by contrast to the mathematical ones, would be inductive.

William Whewell, Charles Sanders Peirce, Henri Poincaré, and above all Karl Popper, criticized inductivism for failing to fit scientific practice. They were right of course, as the characteristic of the advanced sciences is that they contain high-level hypotheses, such as Newton's laws, that go far beyond the relevant empirical data. Moreover, Newton's Principia, *the pinnacle of the Scientific Revolution, started by Galileo and his contemporaries, was a monument to the hypothetic-deductive method.*

Ironically, in Newton's England, Francis Bacon's authority was so great, and rejection of Cartesian apriorism so vehement, that Newton himself declared having proceed inductively. And at the beginning of the nineteenth century, Whewell called 'inductive' the factual sciences, although he was one of the first to point out the importance of non-inductive hypotheses.

The classical positivists, from Comte and Mill to Mach and Ostwald, cultivated inductivism. And right after the First World War, the logical positivists and other scholars started to talk about an inductive logic, parallel to deductive logic, and regarded as an application of the calculus of probability. Nowadays there still is talk of inductive logic, though far less enthusiastically. Is this theory valid, and does it work? Let us see.

12.1 The Semantic Objection to Inductivism

John Dalton recorded meteorological data all his life, in hopes of finding the weather laws by induction. He did not find them, but we honor him for having founded atomic chemistry. This was then a bold speculation that the positivist Comte repudiated just as vehemently as the idealist Hegel.

M. Bunge, *Evaluating Philosophies*, Boston Studies in the Philosophy of Science 295, DOI 10.1007/978-94-007-4408-0_12,

The critics of inductivism, from Poincaré to Popper, have pointed out its poverty, but have not explained why it cannot fly. The reason that it is impossible to ascend from empirical data to high-level law statements is that the latter contain *concepts that do not occur in the empirical evidence for those statements.* For example, "The butler dunit" does not contain the concept of a fingerprint. And "All men are mortal" offers no hint about the molecular and cellular aging mechanisms.

An old scientific example of the data-hypothesis chasm is this: The data of planetary astronomy are distances, angles, and times, but the basic laws of motion also contain the key concepts of mass and force, absent from the data obtained through telescopes. In other words, whereas celestial mechanics is a dynamical theory, the astronomical data are kinematical.

Something similar holds for electrodynamics, atomic physics, genetics, sociology, archaeology, and other fields. For example, the energy levels of a light atom can be calculated but not observed; what can be observed is the light emitted by the atom when it decays from one energy level to a lower one, or that it absorbs while jumping to a higher level. Likewise, hominims are unobservable: only some of their fossil bones, artifacts, and tracks can be observed. And yet, once in a while one encounters the claim that the scientific laws may be obtained from empirical data, and even that such process can be automatized. For example, a recent issue of *Science* contains a paper by the sociologists James Evans and Andrey Rzhetsky (2010), titled "Machine science", where they prophesy that within the next decade algorithms to generate hypotheses from experimental data will be constructed. Regrettably they do no reveal the algorithm employed to generate their hypothesis.

It is not necessary to resort to grand scientific theories to refute inductivism, for problems of the same kind arise in daily life. For example, to understand the behavior of a given person we conjecture her motives. Whenever we wish to jump from effect to cause we tackle an inverse problem, that we attempt to solve by imagining a hypothesis involving predicates that do not occur in the extant data.

What is more, when performing such inferences we invent not only substantive hypotheses but also indicators, which are methodological hypotheses that link unobservables to observables. For example, we suspect that the thief was that bedraggled youth who looks like a drug addict, because we know that drug addicts are always short of cash.

We are usually aware that such "inferences", or correlations between appearances and occult realities, are just as fallible as our substantive hypotheses. And most of us harbor no illusions that we may ever polish and regiment our non-deductive "inferences" to the point of logical rigor. The enthusiasts of inductive logic are more optimistic. Let us peek at this theory.

12.2 Bayesian Induction

The mathematicians Sir Harold Jeffreys (1939) and I. J. Good (1983), the statisticians Bruno De Finetti (1972) and Dennis Lindley (1976), and the philosophers Rudolf Carnap (1950) y Georg Henrik von Wright (1951), as well as their numerous

American followers, tried to introduce rigor in the inductivist program by using the calculus of probability. Most of them worked on what is usually called Bayesian induction, which interprets the probability of a proposition as the measure of its credibility (see Chap. 11).

More precisely, this school handles the antecedent probabilities (or priors) $P(h)$ of a hypothesis h, and $P(e)$ of an empirical datum e relevant to h, as well as the a posteriori, or conditional, probability of h given e. By definition, this probability is $P(h|e) = P(h \,\&\, e)/P(e)$. Regrettably, we are not told how to evaluate these probabilities, other than by gut feeling or intuition.

An original and basic idea of this school is that of inductive support or degree of confirmation. The relevant evidence e is said to *confirm* hypothesis h if the conditional probability of h given e, or $P(h|e)$, is greater than the probability $P(h)$ evaluated before getting e. Further, Bayes's theorem is used to relate $P(h|e)$ with the "inverse" or dual probability $P(e|h)$ of e given h: $P(h|e) = P(e|h).P(h)/P(e)$.

Another basic and attractive idea is that, as research proceeds, it supplies new data, which allow one to update the hypothesis in question. If the successive values of $P(h|e)$ converge to a number close to 1, we will be reassured that our research is on the right track.

In short, the Bayesian inductivists employ the potent machinery of probability theory to deduce an impressive pile of formally rigorous theorems. The question is to know whether the probabilities of propositions make sense, and thus inductive logic is more than an academic industry. Let us tackle this question.

12.3 A Radical Criticism of Bayesian Induction

The critics of inductivism have exhibited some of the flaws of inductive logic. In particular, Popper (1959) showed that the probability of a universal hypothesis is null. However, none of those critics has been radical, since all of them admitted that propositions can be assigned probabilities—even though nobody knows what the probability of a proposition means. In particular, the central idea of Popper's (1963: 218) theory of the verisimilitude of a proposition is that it is the more verisimilar, the more improbable.

None of those critics noted that

1. it makes no sense to attribute probabilities to propositions except in the context of ordinary knowledge, where—as De Finetti (1972) pointed out—the word 'probability' is often used as a synonym of 'credibility';
2. inductive logic does not supply any criteria or rules for evaluating the probabilities in question: these are left to the philosopher's whim;
3. inductive logic presupposes that all propositions can be directly contrasted to empirical data; this is true for empirical generalizations such as "All dogs can bark", but not for high-level hypotheses, which must be combined with indicators—that is, bridges between the hypothetical and the empirical;
4. scientists never evaluate their research projects or their findings assigning them probabilities and checking whether these converge to unity;

5. scientists cherish new truths rather than entrenched beliefs, so that philosophers should work out the concepts of plausibility and truth, leaving those to belief and credibility to cognitive and social psychologists.

12.4 Concluding Remarks

In summary, inductive logic does not help understand, let alone advance, any aspect of scientific research. This is why it is never used to referee grant applications or papers submitted for publication, in particular papers on inductive logic. Moral: Formal rigor and originality are worthless unless combined with clear meaning and relevance.

References

Bunge, Mario. 2006. *Chasing Reality*. Toronto: University of Toronto Press.

Bunge, Mario. 1998. *Philosophy of science*, 2 vols. Updated version of *Scientific Research* (New York: Springer, 1967). New Brunswick: Transaction.

Carnap, Rudolf. 1950. *Logical foundations of probability*. London: Routledge & Kegan Paul.

De Finetti, Bruno. 1972. *Probability, induction, and statistics*. New York: Wiley.

Evans, James, and Andrey Rzhetsky. 2010. Machine science. *Science* 329: 399–400.

Georg Henrik, Von Wright. 1951. *A treatise on induction and probability*. London: Routledge & Kegan Paul.

Good, I.J. 1983. *The foundation of probability and its applications*. Minneapolis: University of Minnesota Press.

Jeffreys, Harold. 1975 [1939]. *Scientific inference*, 3d ed. Cambridge, UK: Cambridge University Press.

Lindley, Dennis. 1976. Bayesian statistics. In *Foundations of probability theory, statistical inference, and statistical theories of science*, vol. 2, ed. W.L. Harper and C.A. Hooker, 353–362. Dordrecht/Boston: D. Reidel.

Popper, Karl R. 1963. *Conjectures and refutations*. New York/London: Basic Books.

Popper, Karl R. 1959 [1934]. *The logic of scientific discovery*. London: Hutchinson.

Chapter 13
Bridging Theories to Data

All philosophers admit the difference between empirical concepts, such as those of color and refraction, and theoretical ones, such as those of wavelength and refraction index. But not all concur on what to do with the gap in question. Empiricists wish to eliminate all theoretical concepts, or at least to reduce them to empirical concepts. By contrast, realists admit the mutual irreducibility of the concepts of both kinds, and they attempt to relate them, in the way the state of the cardiovascular system of a patient is diagnosed by measuring her blood pressure, and one tries to "read the mind" of a person by observing her behavior.

In this chapter we will argue that scientists adopt the realist strategy of bridging the theory-data, or hidden-perceptible gap, by constructing indicators, such as pointer readings. But before doing this we shall see briefly what became of the empiricist attempts to reduce the theoretical to the empirical. In particular, we shall recall the two best known such attempts: the radical, or rejection of everything beyond experience, and the moderate, or reduction of the theoretical to the empirical.

13.1 The Attempt to Replace the Theoretical with the Empirical

Radical empiricism, or phenomenalism, holds either that there are only appearances or phenomena (metaphysical phenomenalism), or that only phenomena can be known (epistemological phenomenalism). Kant and Mach were radical phenomenalists, whereas Hume and Comte were only epistemological phenomenalists. Those four thinkers were the most influential phenomenalists, along with Ptolemy in antiquity and Bridgman in recent times. In particular, Comte and Mach opposed the atomic theory, and thus slowed down the progress of physics.

The program of reducing the theoretical to the empirical was adopted, in particular, by the eminent experimental physicist Percy Bridgman, the influential neo-positivist philosophers Rudolf Carnap and Hans Reichenbach, and the logician William Craig.

M. Bunge, *Evaluating Philosophies*, Boston Studies in the Philosophy of Science 295,
DOI 10.1007/978-94-007-4408-0_13,

Bridgman (1927) proposed his "operational definitions," that allegedly equate scientific concepts with laboratory operations, in particular measurements. For example, time would be what clocks measure, or even what they make.

Operationism is false. First, whereas definition is a conceptual operation, measurement and experiment are empirical. Second, some important concepts, such as those of system, boundary, and reference system, are not quantitative. Third, in principle all magnitudes can be measured in alternative ways. And fourth, all precision measurements are designed with the help of theories. At all events, Bridgman ended up by admitting "pencil and paper" operations, that is, conceptual ones, along with empirical procedures.

Carnap (1936–37) proposed his "reduction sentences" to introduce dispositional terms. For example, one would say that substance *A* is soluble in liquid *B* if and only if, when dropping a sample of *A* in B, one observes that *A* dissolves in *B*. Obviously, this is a confusion between defining and testing. In the sciences the properties of things are attributed to the latter regardless of the empirical operations designed to detect or measure them. This holds even in quantum physics for the properties that things acquire when subjected to experimental perturbations. Finally, Craig (1956) invented a procedure for replacing the "auxiliary expressions" (theoretical terms) of a theory with empirical expressions. His method assumes that the theoretical terms vanish as consequences are deduced from the axioms—which in general is not true. Craig did not offer a single example, and did not realize that his procedure amounts to demolishing theories, not to "cleansing" them (Bunge 1967a).

In short, the program of replacing the theoretical with the empirical failed miserably. It was logically impossible, and it went against the historical current: all the valid empirical generalizations have eventually become consequences of assumptions that are far removed from experience. In principle, all scientific concepts can be absorbed by theories.

Phenomena proper, that is, appearances to some sentient being, do not occur in the latter's external world but in the interface between the two. Galileo, Descartes and Locke knew this. In Kantian words, though not in the spirit of Kant's philosophy, the things in themselves have only primary properties; only the things for us have secondary properties. Hence the corresponding phenomenal predicates, such as "white" and "nice," belong in psychology and the social sciences, not in the natural ones.

The preceding can be summed up in the following argument, which is a sort of dual of Craig's theorem mentioned above:

Two statements are comparable with one another only if they share predicates.	[1]
The theories about natural objects do not contain phenomenal [subject-bound] predicates.	[2]
The empirical protocols contain only phenomenal predicates.	[3]
The conjunction of [1], [2] and [3] implies that theories cannot be directly checked by the relevant data.	[4]

Since the data relevant to a theory are not included in the latter, the theory-data connection must be sought outside the theory. However, neither the above-mentioned authors nor their critics (except for Bunge 1967b) realized that the empirical testing of theories requires enriching them with indicator hypotheses. These bridge the theoretical to the empirical, just as the corresponding indicators link unobservables to observables. Such bridges are particularly evident when designing an experiment with the help of precision measuring instruments.

13.2 Philosophers Rediscover Experiment

Until recently almost all philosophers of science focused their work on theories, and overlooked experiments. Ian Hacking's *Representing and Intervening* (1983) was an instant hit because it told philosophers something that scientists had known since the Scientific Revolution: that experiment trumps observation, and is no less important than conceptual representation.

In 1920 the physicist Norman Campbell (1957) published a frequently quoted book on measurement, which he regarded as the defining characteristic of physics. However, he did not discuss the concept of an indicator, even though it is central to measurement. For example, a spring balance shows positions or angles, not weights. The latter are built into the scale by virtue of Hooke's law, about the proportionality between the spring's stretching and the load. That is, the pointer *indicates* the weight thanks to an indicator hypothesis—Hooke's law. And, like all hypotheses, the indicator ones are fallible. (Hence, operationism notwithstanding, they cannot be used to define magnitudes.)

Four decades later, what Patrick Suppes and his coworkers (e.g., Suppes and Zinnes 1959) called 'measurement theory', attracted the attention of a number of psychologists and philosophers. But it was soon shown (Bunge 1973) that the said theory had fatal flaws. Indeed, it involved the confusion of the mathematical concept of a measure (such as an area) with the empirical procedure of measurement, such as weighing. Moreover, it dealt only with extensive (or additive) magnitudes such as length, while many of the basic magnitudes, such as the various densities, are intensive (non-additive). Nor did the theory elucidate the important concepts of dimension (such as LT^{-1} for velocity) and unit, formalized elsewhere (Bunge 1971). More to the point of this chapter, the theory in question did not involve the concept of an indicator (or symptom, or marker), or unobservable-observable bridge, such as the fever-infection and gravity-pendulum period relations.

Not even van Fraassen's latest book, where measurement has pride of place, mentions indicators: it presupposes that all measurements are direct, like those made by carpenters using measuring tapes. This omission is congruent with the Hume-Kant-Comte-Mach-Duhem-Carnap-Goodman phenomenalism that van Fraassen has been advocating for decades. Indeed, if, as Kant (1787B: 724) held, the universe is the totality of phenomena (appearances), then there is no point in looking for unobservables behind them. (It may be objected that Ernst Mach was at the same time a phenomenalist and an eminent experimentalist and, as such, a conspicuous user of indicators. But this only goes to prove that his philosophy failed to match his science.) Worse, van Fraassen (2008: 116) adopts the operationist principle that the very construction of a magnitude involves reference to its measurement (Bridgman 1927). Indeed, he writes: "The questions *What counts as a measurement of (physical quantity) X?* and *What is (that physical quantity) X?* cannot be answered independently of each other."

But of course it is impossible to design and operate a measuring instrument without having some idea of what it is expected to measure. Second, for many magnitudes there are several very different alternative measurement procedures: just think of distance and time. Third, other magnitudes, notably the potentials, are not measurable at all. (For example, the electromagnetic vector potential A generates the magnetic field intensity H via $H = \nabla \times A$, but not conversely.) Fourth, in a number of cases, such as those of continuum mechanics, statistical mechanics, and atomic physics, theory preceded measurement. Fifth, as will be argued anon, there is no scientific measurement without indicators built into them, for only they allow one to read precision instruments. In short, concept formation precedes empirical test. This is why the reference and sense of concepts must be elucidated with total independence of empirical procedures (Bunge 1974a, b).

Philosophers have largely ignored the earliest analytical writing on observation, measurement, and experiment (Bunge 1967a), perhaps because these involved the concept of an indicator, which is unfamiliar to philosophers. Another reason for that oversight may be that the said book showed that the standard view of the logic of theory testing is mistaken for the same reason: because it overlooks the need for translating the hypothesis to be tested into laboratory or field terms.

For example, the data relevant to electrodynamics do not contain the basic concepts of this theory, those of potential and current density, which are not directly measurable. Likewise, the energy operators and the state functions, basic to the quantum theory, are inaccessible to measurement. In general, the more logically basic a concept, the farther removed from the relevant data it is.

13.3 What Measuring Instruments Show

Measuring instruments are supposed to show, typically either on dials or on digital displays, the "tropes," or values of the desired properties, such as 100 amp. The properties studied by the "hard" sciences are objective or primary, like wavelength, not subjective or secondary, like color. A result of measuring such a property is condensed in a sentence such as "This body is 24 cm long," or "$L(b, \text{cm}) = 24$." for short. But, as relativistic physics has shown, the values of some objective properties depend on the reference frame as well as on the thing in question. Hence, whenever high speeds are involved, the reference frame f must be explicitly mentioned. Furthermore, in the case of precision measurements, the relative error (or standard deviation) σ must be mentioned as well. Thus, an outcome of a length measurement will be of the form "$L(b, f, u) = r \pm \sigma$," where r designates a rational number and u a length unit.

The scale, the ruler and the clepsydra were the earliest measuring tools: in fact, craftsmen and traders used them about five millennia before they entered the laboratory. Of course, modern instruments are far more complex than their ancient precursors. But most measuring instruments, whether ancient or modern, do not show directly the value of a property: what they show instead is the value of an *indicator*, that is, an observable counterpart of the unperceivable item. See the following table.

Factual item	*Indicator*
Atmospheric pressure	Height of barometric column
Wind speed	Angle of anemometer pointer
Magnetic field	Movement of magnetic needle
Electric current intensity	Angle of galvanometer needle
Passage of charged particle	Geiger click
Crystal structure	X-ray diffraction pattern
Recession of galaxies	Galactic redshift
Acidity	pH
Health	Vital signs
Stress	Cortisol level
Economic activity	GDP
Income inequality	Gini index
Quality of life	UN human development index
Probability	Relative frequency

For instance, what one reads in a mercury thermometer is the height of the mercury column. One assumes that the temperature is a function of the height of the mercury column. [More precisely, one uses the hypothesis "$h = h_0(1 + \alpha t + \beta t^2.$")].

Second example: what one sees in a film exposed to the X-rays that went through a crystal are parallel bands or concentric rings. One interprets the figure as results of the diffraction of the X rays by the crystal. One infers the crystal structure (an inverse problem) by solving a bunch of direct problems of this kind: one conjectures some plausible crystal configurations, uses Fourier analysis to calculate the resulting diffraction patterns, and compares them with the one seen in the film. This is of course how the founders of molecular biology tested their conjectures about DNA structure. (More on inverse problems, another subject neglected by most philosophers, in Bunge 2006).

13.4 Visualizing the Unseen

Much the same holds for experimental particle physics. For example, one cannot see proton trajectories: what one can see instead are the tracks left by protons and other charged particles in cloud chambers or on photographic plates. And to "read" these tracks one uses the theory of ionization, that teaches that the dots in the visible trajectory (droplets or exposed emulsion grains) are the more dense, the lower the particle energy—which allows one to find out the direction of motion.

Particle physicists also make frequent use of the law of momentum conservation, which holds in quantum physics as well as in classical physics. This law is particularly handy in interpreting trajectories that seem to originate from nothing or to disappear into nothingness, as in the processes represented by the following diagrams:

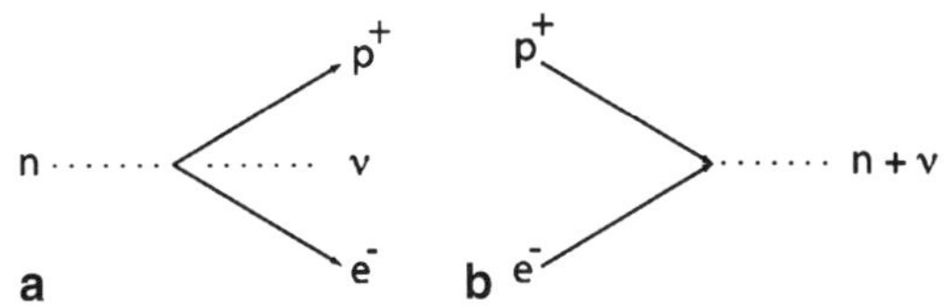

Figure (a) symbolizes the disintegration of a neutron (dotted line) into a proton, an electron, and a neutrino. Figure (b) depicts the nearly inverse reaction: the synthesis of a neutron and a neutrino out of a proton and an electron. The dotted lines stand for invisible (because electrically neutral) particles. To figure out these reactions, physicists use not only knowledge about the incident beams, but also the law of momentum conservation: the vector sum of the momenta of the "visible" particles must equal the resulting momentum of the invisible thing or things. In both cases this law suggests that the missing component, though invisible, is actually there. That is, the presence of the neutral particles is betrayed by its not leaving a trace—just like the silence of the hound of the Baskervilles in Conan Doyle's story suggested to Holmes that the dog was familiar with the criminal. Once more, and contrary to the positivist dogma, data do not cover all the facts out there, whereas good theory does.

A crucial difference between the indicators used in physics, chemistry, and biology, on the one hand, and those used in daily life and in the social sciences on the other, is this. Whereas the former are backed by reliable theories, the latter are either empirical or backed by dubious theories—as when national development used to be identified with GDP growth.

13.5 Theory-Based Indicators

A *theoretically justified indicator* is an observable property v that is functionally related to an unobservable u by a precise formula of the form $u = f(v)$ belonging to a well-confirmed theory. (Roughly, the inverse of f maps states of the mensurandum into states of the mensurans. In other words, f is the code that allows one to read noumena from phenomena.)

An indicator formula is not fundamental, but part of the theory of the measuring instrument in question. For example, the spring balance involves Hooke's law, that the stretching of the spring, shown by the pointer, is proportional to the load. In the case of the galvanometer, electrodynamics shows that the intensity of an electric current is proportional to the tangent of the angle of the deviation of the needle. This suggests that there is a specific theory behind every pointer reading—a fact emphasized by Pierre Duhem (1914) a century ago.

Paradoxically, some measurements do not involve measuring instruments. Two famous cases of this kind are Rutherford's and Millikan's measurements of the electric charge of particles. The former boiled down to measuring the radius of curvature of the particle trajectory under the action of a magnetic field perpendicular to the plane of the particle trajectory. In this case, the theoretical formula that justified the use of such radius as a charge indicator was a consequence of Lorentz's formula for the magnetic force acting on the particle.

The electric part of the very same formula was behind Millikan's measurement. This formula says that, in this case, the downward gravitational pull is counteracted by the upward electric force generated by the electric capacitor. When the observer sees through a microscope that a tiny oil drop between the plates of the condenser has ceased moving up and down, he knows that the two forces are equal, and that he can use the said formula to calculate the total charge on the drop. No wonder Millikan was rewarded with the Nobel prize for imagining this measurement, at once path-breaking, simple, and beautiful. However, it was his lab assistant who came up with the idea of using a barber's sprayer to produce a fine spray of electrically charged oil drops.

Presumably, the Sumerian craftsmen and traders who invented or used the earliest measuring devices did not lose much sleep over scientific indicators: they proceeded empirically. On the other hand, the nineteenth century scientists and engineers who designed the earliest precision instruments must have invested much ingenuity in mapping theoretical unobservables into laboratory observables. Some of that ingenuity went into writing the laboratory manuals familiar to physics students, such as Kohlrausch's (1873) classical textbook on physical measurements.

The fact that the design of good experiments is oriented by theories explains the utter failure of the pedagogical constructivists—who claim that children will find the laws of nature by themselves—to inspire and educate future scientists. Besides, theories are invented, not discovered; and all scientific theories contain concepts, such as those of atom, metabolism, social cohesion, and price elasticity, that denote imperceptible items.

The same facts explain the impotence of amateurs in the laboratory, where all they can do is to stand in the way of researchers and break pieces of equipment. This is why the sociologists and anthropologists who, like Latour and Woolgar (1979), spent some time in a laboratory, were unable to understand what their subjects were up to. Consequently they gave an utterly distorted description of scientific research, namely as "social construction" of reality and struggle for power. (Think of Galileo and his fellow astronomers as constructing the sunspots and attempting to unseat the Pope.)

Some sociologists in the late twentieth century realized that the construction and empirical test of quantitative hypotheses requires indicators. They started what became known as the social indicators movement, which in 1974 got a journal of its own, *Social Indicators Research*. The same year UNESCO held a conference on development indicators, where a multidimensional indicator of human development was first suggested (Bunge 1974c). This was a precursor of the three-dimensional (biological, economic, and educational) indicator of human development adopted by the UN in 1989 despite the resistance of the economists who claimed that the GDP, a measure of economic activity, was sufficient.

13.6 The Place of Indicators in Theory Testing

We are now ready to look at the place of indicators and indicator hypotheses in the experimental testing of scientific theories. What can be empirically tested are not general theories, such as classical or quantum mechanics, but theoretical models, or applications of such theories, to special cases, such as a theory of a planet or of a helium atom. The reason should be obvious: there is no general body or general atom. Any such model *M* is either built from scratch ("free model") or gotten by enriching the general theory *G* with a set *S* of subsidiary assumptions representing some salient properties of the thing in question. That is, $G \cup S \mid - M$.

But, contrary to the standard view on testing, not even such specialization brings the theory down to the empirical level. To take a theory into the laboratory we must combine the theoretical model *M* with a set *I* of pertinent indicator hypotheses. For example, if the theoretical model to be tested involves the intensity *i* of an electric current, we replace *i* with $k.tan\ \theta$, where *k* is a constant that characterizes the instrument, and the angle θ is read on the dial.

Likewise, the mass spectrograph allows atomic physicists to visualize the differences between the masses of isotopes; chromatography revolutionized organic chemistry by allowing chemists to compare molecular masses through the straightforward expedient of comparing the tracks left on blotting paper by solutions of different compounds traveling under the action of an electric field; and fMRI (functional magnetic resonance imaging) has greatly helped cognitive neuroscience by showing on a screen the metabolic activity of a brain region, an indicator of neural activity. In all these cases the indicator hypotheses bridge unobservables to observables.

In short, what is being directly confronted with empirical data is not a general theory *G*, nor even a theoretical model *M* (or special theory) based on *G*, but its operationalization *O*, which is the translation of *M* into the relevant indicator hypotheses *I* (Bunge 1967a, b). Thus, the process of theory testing involves the construction of a deductive tree of this form:

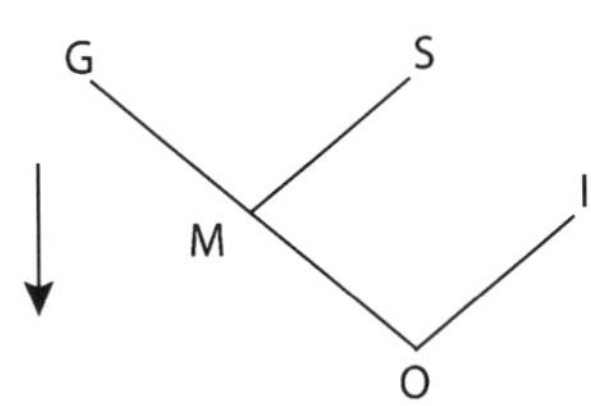

13.7 Indicators in Quantum Physics

Remove indicators, and no empirical tests become possible. And yet, ironically, none of the philosophical champions of empirical testing has ever mentioned indicators—which may be taken as an indicator of the remoteness of their opinions from scientific practice.

However, in this regard the founders of quantum mechanics too sinned. In fact, one of their main dogmas was that, as decreed by what used to be called "the bible of quantum mechanics" (Dirac 1958: 36), "every eigenvalue [of an operator representing a dynamical variable] is the possible result of the measurement of the dynamical variable for some state of the system"—regardless of the measurement procedure. (In 1941 Heisenberg realized that actually quantum mechanics does not comply with the operationist philosophy. This motivated him to propose his *S*-matrix theory, which contained only measurable inputs and outputs. But this theory was abandoned because it did not solve any of the extant problems. Its author did not even mention it in his autobiography.)

Blinded by logical positivism—the dominant philosophy in the physics community at the time—those eminent theorists overlooked the fact that microphysical measurements are extremely indirect, as they involve indicators and pieces of apparatus, such as amplifiers, which only a (classical or semi-classical) theory can explain.

To his credit, Niels Bohr (1958), the father of the (Copenhagen) interpretation in question, rightly emphasized the dependence of the tests of the quantum theory upon classical physics. But he did not realize that any realistic descriptions of measurements, such as his own, are inconsistent with that interpretation of the quantum theory.

In fact, according to the Copenhagen orthodoxy, this theory describes observations rather than independently existing physical objects. Thus, Heisenberg (1969:171) held that atoms "are parts of observation situations." But this claim is shown to be false by remembering that stars, where observations are impossible, are made of atoms. Besides, an analysis of the variables occurring in the basic formulas of the theory shows that none of them involves measuring instruments or even indicators—both of which belong in the laboratory (Bunge 1967b). If theories did contain such items, it would be impossible to confront them with empirical findings: just as judges are not expected to try themselves, so theories are not expected to tell us how to test them.

Unsurprisingly, the general quantum theory of measurement, founded in 1932 by the great mathematician John von Neumann (1955), does not include any indicators. Moreover, this theory is supposed to hold for measurements of all kinds, invasive like those involving colliders, and non-invasive like those using spectrographs. It thus assumes explicitly that quantum mechanics is universal and self-sufficient; and it assumes tacitly the existence of a universal meter. (See Wheeler and Zurek 1983.) But both assumptions are false.

Firstly, quantum mechanics cannot possibly be used to describe macrophysical objects, such as measuring devices, composed of zillions of atoms (e.g., Zeilinger 1990). No one even knows how to write the Schrödinger equation for mesoscopic objects, such as a pollen particle or a DNA molecule. Hence the fashionable claim that the quantum theory holds for all things, regardless of their complexity, is just a dream. True, Stephen Hawking wrote a Ψ purporting to designate the wave function of the universe; but, since he did not specify the corresponding wave equation, it was just bluff.

Secondly, there can be no universal meter, for the simple reason that there is far more than one magnitude: all measuring instruments are specific. For example, chronometers cannot be used to measure masses, and pH-meters cannot double as scales. The design of every precision instrument involves special theories including "constitutive equations," such as Ohm's law. These are law statements about materials of particular kinds, such as metals or photographic emulsions. For example, interferometers work with light beams, not atomic beams; this is why their design and operation requires optics, not gravitation theory or thermodynamics.

Because the general quantum theory of measurement refers to the non-existing universal meter, it is untestable in principle, hence beyond the ken of science. No wonder then that, after eight decades, the general quantum theory of measurement has never been put to the test, and has never helped design any real experiment. In other words, the orthodox view of measurement is superficial: it "has little, if any, relations to the workings of actual physical measurements" (Schlosshauer 2007: 334). It is just an academic industry employing mathematicians and philosophers unfamiliar with laboratory work.

13.8 Concluding Remarks

Experimental design, instrumentation and measurement require theoretical inputs of two kinds: substantive and instrumental. The former is about the object measured, and it alone can tell us *what* to measure, though not *how* to measure it. Only a special theory of the measuring device can tell us how it works, in particular whether it can help measure the desired property. Without some such knowledge it would be impossible to design the instrument in question. Imagine the Reverend Atwood trying to design his accelerometer without Newton's second law of motion, or Rutherford doing nuclear physics without Lorentz's classical equation of motion for charged particles.

Again, only theoretical particle physics tells the experimenter that the charge/mass ratio is worth measuring; and the wave theories of light show that wavelength measurements are central. But it is up to the experimenter's ingenuity to devise means to perform the measurements in question. Thus, none of the quantum theorists devised any of the particle colliders, from the cyclotron onwards. But none of these experimental breakthroughs would have been possible in the absence of the relevant substantive theories. Substantive theories pose laboratory problems, and instrumental theories solve them.

The preceding suggests a few philosophical morals. First: Measuring instruments, together with the indicators they embody, bridge imperceptible noumena (things-in-themselves) to the corresponding phenomena (things-for-us). Second: The mere existence of such bridges falsifies phenomenalism, the doctrine according to which noumena either do not exist or, if they do, are unknowable. Third: The fact that, to be tested, a high-level hypothesis has got to be conjoined with indicator hypotheses, falsifies the naïve logic of empirical testing held by positivists and Popperians alike.

In other words, when measuring an unobservable property we construct a chain like this

Object	$\xrightarrow{I}$ *Apparatus*	$\xrightarrow{Signal}$ *Brain*
Unobservable	Observable	Observation

Thus, contrary to the Humean, Kantian, positivist, conventionalist, and Copenhagen doctrines, phenomena occur at the end of the chain, not at its beginning, since they can only happen in a living brain. The latter, though, is in control of every link in the artificial chain: it chooses and prepares the object, designs and operates the apparatus, and reads and interprets the latter's result. (While some of these steps can be automated, only living brains can invent original measuring devices and automate their operation.)

In sum, it has been known since the 1600s that experimental interventions suggest or test representations of reality. But all experimental designs are based on theories—a case of virtuous circle. No intervention without representation.

References

Bohr, Niels. 1958. *Atomic physics and human knowledge*. New York: Wiley.

Bridgman, P.W. 1927. *The logic of modern physics*. New York: Macmillan.

Bunge, Mario. 1967a. *Scientific research*, 2 vols. New York: Springer. Rev. ed: *Philosophy of Science*. (New Brunswick: Transaction 1988).

Bunge, Mario. 1967b. *Foundations of physics*. New York: Springer.

Bunge, Mario. 1971. A mathematical theory of the dimensions and units of physical quantities. In *Problems in the foundations of physics*, ed. M. Bunge, 1–16. New York: Springer.

Bunge, Mario. 1973. On confusing 'measure' with 'measurement' in the methodology of behavioral science. In *The methodological unity of science*, ed. M. Bunge, 105–122. Dordrecht: Reidel.

Bunge, Mario. 1974a. *Sense and reference*, Treatise on basic philosophy. Dordrecht: Reidel.

Bunge, Mario. 1974b. *Interpretation and truth*, Treatise on basic philosophy. Dordrecht: Reidel.

Bunge, Mario. 1974c. The methodology of development indicators. UNESCO, Methods and Analysis Division, Dept. of Social Sciences, 1974. Rev. version: Development indicators. *Social Indicators Research* 9: 369–385, 1981.

Bunge, Mario. 2006. *Chasing reality*. Toronto: University of Toronto Press.
Carnap, Rudolf. 1936 – 37. Testability and meaning, In *Philosophy of Science*, 3: 419–71, 4: 1–40.
Campbell, Norman Robert. 1957 [1920]. *Foundations of science*. New York: Dover.
Craig, William. 1956. *Replacement of auxiliary expressions*, Philosophical Review, 65: 38–55.
Dirac, P.A.M. 1958. *The principles of quantum mechanics*, 4th ed. Oxford: Clarendon.
Duhem, Pierre. 1914. *La théorie physique*, 2nd ed. Paris: Rivière.
Hacking, Ian. 1983. *Representing and intervening*. Chicago: University of Chicago Press.
Heisenberg, Werner. 1969. *Der Teil und das Ganze*. München: R. Piper.
Kant, Immanuel. 1787. *Kritik der reinen Vernunft*, 2nd ed. Hamburg: Felix Meiner.
Kohlrausch, Friedrich. 1873. *Introduction to physical measurements*. London: J. & A. Churchill.
Latour, Bruno, and Steven Woolgar. 1979. *Laboratory life: The social construction of scientific facts*. London/Beverly Hills: Sage.
Schlosshauer, Maximilian. 2007. *Decoherence and the quantum-to-classical transition*. Berlin/Heidelberg/New York: Springer.
Suppes, Patrick, and Joseph L. Zinnes. 1963. Basic measurement theory. In *Handbook of mathematical psychology*, vol. I, ed. R.D. Luce, R.R. Bush, and E. Galantner, 1–76. New York/London: John.
Van Fraassen, Bas C. 2008. *Scientific representation: Paradoxes of perspective*. Oxford: Clarendon.
Von Neumann, John. 1955 [1932]. *Mathematical foundations of quantum mechanics*. Princeton: Princeton University Press.
Wheeler, John A., and Wojciech H. Zurek (eds.). 1983. *Quantum theory and measurement*. Princeton: Princeton University Press.
Zeilinger, A. 1990. Experiment and quantum measurement theory. In *Quantum theory without reduction*, ed. Cini Marcello and Jean-Marc Lévy-Leblond, 9–26. Bristol: Adam Hilger.

Chapter 14
Matter and Energy: Physical or Metaphysical Concepts?

All the sciences and technologies of reality deal with concrete or material things endowed with energy, and the materialist worldview emerged in Greece and India two and a half millennia ago, and flourished in the French Enlightenment. And yet, there is still no consensus about the definitions of the general concepts of matter and energy. Let us deal with both of them, starting with the second.

14.1 Energy: Stuff or Property?

The word *energy* has an ancient lineage, but it was ambiguous until mid-nineteenth century, when the great polymath Hermann von Helmholtz formulated the principle of conservation of energy. Even then, and for the duration of the century, no clear concept of energy was available.

Around 1900 all physicists and chemists regarded energy, correctly, as a property of material thing, not as a substance. For example, they wrote formulas such as $E_k(x) = (1/2)mv^2$ for the kinetic energy of a particle x with mass m and velocity v. Another famous formula is the one that relates the electromagnetic energy E of a photon with its frequency ν, namely $E = h\nu$. In this formula, E and ν stand for properties of a material thing, an arbitrary photon, whereas h, Planck's constant, denotes neither thing nor property.

Only a few scientists, notably the eminent physical chemist Wilhelm Ostwald, thought that energy is not a property but a substance, no less than the fundamental stuff. Strangely, also Niels Bohr, one of the fathers of quantum physics, wrote once that energy is "the primary substance of the world".

Ostwald and a few others went as far as to try and build energetics, the universal science that would study everything, from stones to values, and would erase the distinction between matter and mind, thus overcoming the materialism/idealism dualism. This program was never fulfilled. It could not be fulfilled because there is no such thing as energy in itself: every energy that is measured, calculated or

M. Bunge, *Evaluating Philosophies*, Boston Studies in the Philosophy of Science 295, DOI 10.1007/978-94-007-4408-0_14, © Springer Science+Business Media Dordrecht 2012

conjectured is the energy *of* a material object, whether body, particle or field. Let us now jump to the present day.

14.2 Special Energies

The word *energy* designates of course a key concept in modern physics, chemistry, and engineering. And yet, in the famed Feynman *Lectures on Physics* we read that present-day physics does not know what energy is. Is this true? Let us see.

It is well known that there are various kinds of energy: kinetic, elastic, thermal, gravitational, electric, magnetic, nuclear, chemical, etc. More precisely, there are as many kinds of energy as kinds of process. Mayer, Joule y Helmholtz exaggerated somewhat when they wrote about "conservation of energy" in general. Actually they dealt only with two kinds of energy: mechanical and thermal. In particular, Joule's beautiful experiment proved just the quantitative equivalence between kinetic energy and thermal energy.

Unlike other species, the various kinds of energy are equivalent in that they can be transformed into one another. For example, when we tense a bow we store potential elastic energy in it; and when we release the bow, that energy is transformed into the arrow's kinetic energy. In this two-step process, the type or quality of energy changes (from potential to actual), but its quantity remains the same.

Such quantitative conservation is the reason that we regard all the kinds of energy as mutually equivalent. In other words, the introduction of the general concept of energy is justified by the general principle of conservation of energy. But, paradoxically, both the concept and the principle, though firmly rooted in physics, overflow it. Let me explain.

Every particular concept of energy is defined in a given chapter of physics. For example, kinetic energy is defined in dynamics; thermal energy in thermodynamics; electromagnetic energy in electrodynamics; nuclear energy in nuclear physics, and so on. Every one of these fields handles its own concept of energy. Moreover, in all of these fields, save statics, thermostatics, electrostatics, magnetostatics, and geometrical optics, a theorem of conservation of energy is provable from the corresponding equations of motion or field equations.

In the above-mentioned exceptions one feigns that nothing happens—for example, that the light rays have always been there, rather than being the trajectories of something that moves. In these cases the concept of energy plays no role because that of becoming does not occur either. This suggests an equivalence to be introduced below: Energy if and only if becoming.

To add energies of two or more types we need to join the corresponding disciplines. This is achieved by adding them a "bridge formula" containing concepts of all the disciplines in question. For example, the total energy of a jet of an electrically charged fluid can only be calculated in the hybrid science of electro-magneto-thermo-hydrodynamics, a key discipline in the study of stellar atmospheres.

¿What is the range of the energy concept, that is, things of what kinds can energy be attributed? A mathematical physicist might be tempted to answer: it may be attributed to all the things describable by a hamiltonian theory. To be sure, one may postulate Hamilton's canonical equations for an unspecified concrete thing, from electron to business firm. But one cannot claim that the hamiltonian function or operator occurring in them represents the energy of the thing unless the generalized coordinates and momenta occurring in it are interpreted in physical terms, And in turn this requires specifying the kind of thing that is being referred to—which restricts the type of energy.

All the sciences that study concrete or material things, from physics to biology to social science, use one or more concepts of energy. For example, cognitive neuroscientists want to measure the metabolic cost (in calories) of a bit of information transmitted across a synapse; anthropologists, sociologists and economists are interested in finding out the energy consumption per capita, and whether people work so as to optimize their efficiency, or ratio of energy output to energy input.

14.3 Minitheory

Because it is ubiquitous, the concept of energy must be philosophical and, in particular, metaphysical (or ontological). That is, it belongs in the same league as the concepts of thing and property, event and process, space and time, causation and chance, law and trend, and many others.

Assuming that the general concept of energy is indeed philosophical, let us proceed to analyzing it in philosophical terms. And, since the best analysis is synthesis, or the embedding into a theory, let us construct a mini theory centered in the concept of energy.

We start by identifying energy with changeability. That is, we propose the following

Definition Energy = changeability

This convention may be rewritten as follows:

For all x: (x has energy = x is changeable).

Let us now put this definition to work. We begin by assuming

Postulate 14.1 All concrete (material) objects, and only they, are changeable.
That is,
For all x: x is concrete (material) if and only if x is changeable.
In logical symbols, $\forall x\,(Mx \Leftrightarrow Cx)$.

Remark 14.1 We have equated "concrete" with "material". This convention is more common in philosophy than in physics. According to it, fields are just as material as bodies. For example, photons are material in the philosophical sense of the word although they lack mass, solidity and a shape of their own—the attributes of matter before the emergence of field physics in mid-nineteenth century.

The above Definition and Postulate 14.1 entail

Theorem *For all x: if x is a material object, then x has energy and vice versa.*

$$\text{Shorter}: \forall x(Mx \Leftrightarrow Ex).$$

This theorem has two immediate consequences. The first is

Corollary 14.1 Abstract (non-concrete) objects lack energy.

That is, as Plato had correctly intuited, the conceptual (or ideal) objects are immutable. For example, the concept of energy has no energy. True, it is usually said that the concept of energy has changed in the course of history. Actually, what have changed are the brains that have thought them. Every one of those concepts is intemporal. If preferred, the successive conceptual creations do not change by themselves. This is why we cannot write down equations of motion for concepts.

What happens with the divinity concepts in the innumerable religions? The believers of most of these conceive of their divinities as spiritual entities, hence as devoid of energy. If Postulate 14.1 is adopted, it turns out hat all gods are immutable. To a hylorealist (ontological materialist and epistemological realist), this confirms the atheistic hypothesis that there are no gods. But of course this result won't disturb the sleep of any religious believers: it should only remind them that they must adopt a spiritualist philosophy and stand in guard against all attempts to bring religion close to physics.

The second immediate consequence of the above theorem is

Corollary 14.2 Energy is a property, not a thing, state, or process.

Remark 14.2 Because energy is a property, it can be represented by either a function or an operator. In classical physics one may say that $E(c, x, t, f, u)$ is an arbitrary value of the energy of a thing c situated at point x, and time t, relative to a frame f, and reckoned or measured in the energy unit u. The function in question has the general form

$$E : C \times E^3 \times T \times F \times U \to \mathrm{R},$$

where C is a set of concrete things, E^3 the Euclidean space, T the set of instants, F a set of reference frames, U a set of energy units, and R that of real numbers. In the case of an interaction energy, C will be replaced with the set of pairs <thing, thing>, that is, the Cartesian product $C \times C$. In quantum physics, energy is represented by the hamiltonian operator H. The corresponding property is the energy density $\psi^* H \psi$, which has the same form as the classical energy function.

Remark 14.3 The kinetic energy of a particle, relative to a reference frame bound to the latter, is nil. Similarly, the total energy of a thing embedded in a field becomes zero when its kinetic energy equals its potential energy (as is the case with the outer electron of an atom at the time of ionization). However, null energy is not the same as lack of energy, just as zero temperature (on some scale) is not the same as lack of temperature. In these cases, unlike the case of the photon mass, zero is just a special numerical value. Analogy: the balance of a cow's bank account can never be zero, because cows cannot have bank accounts.

Remark 14.4 Corollary 14.2 entails that the concept of a concrete or material thing cannot be replaced with that of energy. There is no such thing as energy in itself, just as there is no age in itself: every energy is the energy of some thing. Hence energetism, which a century ago was proposed as an alternative to both materialism and idealism, was radically mistaken. However, the energetists, particularly the famous physical chemist Wilhelm Ostwald, were right in holding that energy is universal—a sort of cross-disciplinary currency. They would have been even more right had they proposed.

Postulate 14.2 Energy is the universal physical property: the only property common to all material things.

Remark 14.5 It is generally believed that position in spacetime is another universal physical property. So it is. But, according to any relational (as opposed to absolute) theory of spacetime, the latter is not basic but derived: it is the basic structure of the collection of all things, every one if which possesses energy. Roughly, space is rooted to the spacing of things, and time to their change. No space without things, and no time without change—just as Aristotle thought. If an omnipotent creator decided to annihilate all the material things, spacetime would disappear along with them.

Remark 14.6 This relational conception of spacetime is confirmed by Einstein's theory of gravitation, according to which the geometry of the universe depends upon its content. Indeed, the core of this theory is the formula "$Gab = Tab$", where the Gab describe spacetime, and the Tab matter. These ten equations are incompatible with John A. Weeler's hypothesis, that matter is reducible to spacetime. What is true is that the equations in question allow one to determine the distribution of matter in the universe by performing geometric measurements.

Remark 14.7 Postulate 14.2 does not state that every thing has a precise energy value at any given time and relative to any given reference frame. It does not because, in quantum theory, a sharp energy value is the exception rather than the rule. Indeed, in general a quanton (quantum-mechanical thing) is in a superposition (or weighted sum) of infinity many energy eigenfunctions. And the corresponding eigenvalues scatter around some average values (such as an atomic energy level).

That superposition of functions shrinks to one when the quanton interacts with a macrophysical entity; when this happens the energy distribution shrinks to a single value. This transition is usually called 'decoherence'. (More on this in Chap 15.)

Our final assumption is:

Postulate 14.3 The total average energy of an isolated concrete object does not change in the course of time.

Remark 14.8 This is of course the general principle of conservation of energy. (The rider 'average' is necessitated by the quantum theory.) This principle is so extremely general, that it belongs in philosophy rather than in physics.

Remark 14.9 In an expanding universe, energy is not conserved, but is slightly dissipated.

Remark 14.10 According to quantum electrodynamics, the vacuum energy is not zero but fluctuates irregularly (randomly) around zero. This result of the most accurate of all scientific theories does not invalidate the characterization of energy as the universal property of all concrete things. All it does is to restrict the domain of validity of the classical definition of vacuum as the absence of matter endowed with mass. The electromagnetic field that remains in a region of space after all the electric charges have been neutralized, and all the electric currents have been switched off, is a concrete though tenuous thing. It is so concrete, that it exerts a force upon the electrons of the atoms: this is the Lamb shift, one of several measurable effects of the vacuum. Another is the Casimir force of attraction between two metallic plates in vacuum.

This concludes the presentation of our minitheory of matter and energy. Let us now use it to expose a bunch of common errors.

14.4 Common Errors

New Age scribblers have no monopoly on nonsense about matter and energy. Careless physicists have produced some such nonsense. In fact, energy is often confused with radiation, and matter with mass. Let us examine a small sample of mistakes of this kind in physics textbooks and popular science publications.

Example 14.1 The expression "annihilation of matter" and "materialization of energy", for pair destruction and formation respectively, are incorrect. What gets annihilated when a pair of electrons of opposite charges gets transmuted into two photons is not matter but mass, a property of particles and bodies but not of photons. Dually, matter endowed with mass emerges in "pair creation". (The two sides in the conservation equation, $2mc^2 = hv$ do not hold at the same time. The LHS holds before, and the RHS after the "annihilation".) Unlike energy, mass is not a universal property.

Example 14.2 When an antenna radiates electromagnetic waves for a total of E ergs, it loses E/c^2 grams of mass. This does not exemplify the conversion of matter into energy, but the conversion of part of the energy of the antenna (and the field around it) into radiation energy, with the concomitant decrease in mass. The total energy is conserved.

Example 14.3 The formula "$E = mc^2$" does not "say" that mass and energy are the same modulo c^2. Indeed, E and m are very different properties. For one thing, the former is the ability to change, whereas the second is inertia, or the disposition to resist change in one respect (state of motion). For another, the formula only holds for things endowed with mass, that is, particles and bodies. So much so, that it is a theorem in relativistic mechanics, not in any field theory. Hence it does not hold for photons or gravitons (assuming that these exist).

Example 14.4 It has been said that "$E = mc^2$" proves that physics has dematerialized matter. This claim involves two confusions: the identification of "matter" and "mass", and the belief that energy is a thing, while actually it is a property of material things: there is no energy without things, just as there are no areas without surfaces.

14.5 The Ontological Square

The preceding may be summarized as follows: The predicates "is material", "has energy" and "is changeable" are coextensive (apply to the same objects) even though they are not cointensive (do not have the same connotation). Furthermore, in a materialist perspective, "is real" too is coextensive with "is material."

In sum, in our view all four concepts in question are coextensive, which is why they are sometimes confused. Furthermore, they are logically interrelated. In fact, they constitute what may be called the *ontological square*:

$$\begin{array}{ccc} \textit{Changeability} & \rightarrow & \textit{Energy} \\ \uparrow & & \downarrow \\ \textit{Reality} & \leftarrow & \textit{Matter} \end{array}$$

14.6 Conclusion

We are finally in a position to answer the question that triggered this chapter, namely whether it is true that in physics one does not know what energy or matter are. This is partially true. Indeed, in each chapter of physics we can identify a special type of energy, and state a particular energy conservation law. The general concepts of energy and matter, as well as the general law of conservation of energy, belong in philosophy. However, philosophy cannot elucidate either without the help of physics. It is only by joining the two fields that a precise answer can be given to the question in hand, namely this. Energy is the universal property of matter: the property of being capable of changing in some respect.

What complicates the problem even further, and on occasion misleads even the specialist, is that (a) there are as many kinds of energy as large classes (genera) of process; (b) there are as many concepts of energy as general physical theories; (c) the general concept of energy is so general, that it belongs in metaphysics (ontology);

and (d) consequently the general principle of conservation of energy is philosophical, though rooted in physics. The concept of matter is parallel.

Moral. Physics cannot dispense with philosophy, just as the latter does not advance if it ignores physics and the other sciences. In other words, science and sound (i.e., scientific) philosophy overlaps partially and consequently they can interact fruitfully. Without philosophy, science loses in depth; and without science, philosophy stagnates and becomes irrelevant.

Chapter 15
Does Quantum Physics Refute Realism, Materialism and Determinism?

*It is generally believed that quantum physics refutes realism, materialism, determinism, and perhaps even rationality. These beliefs, central to the so-called Copenhagen interpretation, were held by the very fathers of the new physics, particularly Niels Bohr (*1934*), Max Born (*1953*), Werner Heisenberg (*1958*), and Wolfgang Pauli (*1961*).*

The philosophical kernel of the said interpretation of the mathematical formalism of the quantum theory is the thesis that this theory describes observations rather than natural objects; and that things, at least on the microphysical level, are effects of observations, which in turn would be products of mental events that would escape natural laws. Let us evaluate these statements, which evoke what George Berkeley's had claimed two centuries earlier: that "to be is to perceive or to be perceived."

15.1 The Spiritualist View

The founders of quantum physics and their faithful followers denied emphatically that atoms and the other microphysical entities exist independently of the observer. Thus the physicist David Mermin (1981) asserted confidently that the Moon does not exist as long as nobody looks at it. Another physicist, Richard Conn Henry (2005), went even farther: without offering a single argument, he wrote in the prestigious journal *Nature* that the universe is not material, not even a set of observations: it is mental, just as the *Vedas*, the Hindu scriptures, had stated three millennia ago.

Furthermore, some cosmologists have written about the "participatory" universe, created at least in part by the ubiquitous observer. And the philosopher Galen Strawson (2008) claimed something similar: that all the things that physicists and chemists study are *mind stuff*, so that "real materialism" (his own brand of idealism) implies panpsychism—the academic name of animism.

M. Bunge, *Evaluating Philosophies*, Boston Studies in the Philosophy of Science 295,
DOI 10.1007/978-94-007-4408-0_15,

Neither of these unwitting Berkeleyans (or Kantians) noticed that, if the universe were mental, there would be no difference between fact and fiction; humankind would have preexisted the universe; and psychologists, not cosmologists, would study the cosmos—or perhaps cosmologists would double as psychologists and psychiatrists. Nor did they realize that, if the universe were mental, it would think itself. The philosopher Jesús Mosterín (2006: 401) came close, as he has held that "the universe thinks itself through us." However, there is even worse: the physicist Roland Omnès (1999: 245) claimed that "the sacred is everywhere in the universe and nothing is completely profane." No evidence proffered.

The most remarkable feature about these extraordinary opinions is not that they are mere reeditions of the most primitive of worldviews, namely animism, which we thought had been conclusively refuted by the pre-Socratic philosophers. The most amazing is that those extravagant opinions are proposed nowadays *ex cathedra*, without any solid arguments, just the way Schelling and Schopenhauer did nearly two centuries earlier.

15.2 Resorting to Experiment

The so-called *observer effect* consists in the claim that the observers do not find and study things endowed with certain properties, but create the things they naively believe to be observing, or at least create its properties. This view would condense the typical measurements and experiments in microphysics. Let us see.

By definition, an observation consists of finding out something about a preexisting thing. To be sure, some observations alter to some extent some of the properties of the observed object. For example, when inserting a thermometer in a liquid body, we warm it up or cool it down somewhat; and the light used to localize an electron alters its position. But in both cases the observed object preexists the observation act.

Things change in chemistry and in high-energy physics, particularly in a collider. In these cases new things emerge that did not previously exist in the same place. But these new things, the reaction products, do not pop out of nothing or from the experimenter's mind: they are products of interactions among preexisting things. So much so, that the experimenter attempts to set up and describe the reactions in question.

For example, when colliding with protons (the nuclei of hydrogen atoms), a beam of protons generates positive and negative pions in accordance with the reaction schema $p+p \rightarrow p+p+\pi^{+}+\pi^{-}$. The product consists of preexisting things (protons p) along with qualitatively new things (pions π). This reaction also occurs naturally, as was discovered in 1947 when photographic plates were exposed to cosmic rays at great heights.

The following diagram summarizes the three processes discussed in the preceding lines: (a) passive observation (e.g., astronomical); (b) experiment (realist version); and (c) experiment (idealist version).

Observer	*Experimenter*	*Experimenter*
↑	↑↓	↓
Apparatus	*Apparatus*	*Apparatus*
↑	↑↓	↓
Object	*Object*	*Object*
a	**b**	**c**

In summary, as Lucretius stated two millennia ago, there are *transformations* of some things into other things, but not *ex nihilo* or *ex mente* creations of concrete things. Of course a brain may new create ideas, such as experimental designs and theories. But, however original, ideation is a process involving preexisting ideas. An utterly original idea, one without roots, would be incomprehensible.

For example, astrophysicists know very little about "dark matter," beyond that its constituents are material (changeable and interacting) things that do not emit light. If dark matter were mental, like fairy tales and mathematical theorems, it would be well known. However, let us go back to quantum physics.

15.3 Criticism of the Subjectivist Interpretation

Almost all philosophers, in particular the logical positivists, adopted the Copenhagen interpretation. It is not easy to doubt the word of all those great scientists, even though three eminent physicists—Max Planck, Albert Einstein, and Louis de Broglie—never accepted that doctrine. In fact, all three Nobel Prize winners were consistent realists.

However, critical philosophers won't use the argument from authority, favored by theologians: they will examine the theory in question before lending their support to this or that interpretation of its mathematical formalism. This is what we did earlier (Bunge 1959a, b, 1967, 1973, 1985, 1979, 2010). An updated version of my arguments in favor of realism follows.

Let us start by recalling the gist of realism. It consists of two theses. One is the ontological thesis that the universe exists independently of the knowing subject or observer (ontological realism). The other thesis is the epistemological postulate that a subject may get to know the universe objectively, or in a third-person manner, though only gradually (see Mahner 2001).

Here are two arguments in favor of realism. First: Whoever endeavors to investigate something may doubt the real existence of anything in particular, but not that of his or her entire environment, in particular his laboratory and the instruments housed in it. For instance, if an experimentalist wishes to ascertain whether a hypothesized object X exists outside her mind, she will design an experiment showing that X alters the state of a second object Y of whose existence there can be no doubt, under penalty of falling into so radical a skepticism that it would render the investigation pointless.

Second: All knowers are born and raised in an environment that preexista them, and which is the product of a natural and social evolution that has been going on for millions of years. The views of the subjectivists philosophers, like Kant, that the universe is a set of appearances (to someone), and that space and time only exist in the subject's mind, are inconsistent not only with physics but also with everything we know about the ontogeny (individual development) and the philogeny (evolution) of humans.

In summary, the world preexists the subject, and people can annihilate themselves but not the universe. Moreover, the climatologists and ecologists know well that nature would gain by the extinction of our species. The subjectivist philosophers may be regarded as refined and deep in a philosophy department, but outside such a bubble he will be regarded as an excentric or even as deranged. For example, he won't be able to leave shops without paying for the merchandise he took, alleging that it was his minds' creation.

15.4 Indeterminacy and Decoherence

In classical physics all properties are sharp: so many kilograms, so many kilometers per hour, and so on. By contrast, quantum magnitudes do not always have precise values. For instance, most of the time an electron is not at a definite point in space, but inside a small cube with a certain probability, inside an adjoining little cube with another probability, and so on. That is, the electron has a position distribution—if preferred, it is in a superposition (linear combination) of positions.

The same holds for the remaining dynamical variables of quantum objects: linear momentum, angular momentum, spin, and energy. For example, the energy state of a quantum object is ordinarily a superposition of elementary energy states (eigenstates), every one of which corresponds to a precise value of energy.

Only time, mass, and electric charge lack an intrinsic dispersion—usually called "indeterminacy," or "uncertainty." So do all the inherently macrophysical variables, such as stress, temperature, and entropy: all of these have sharp values at any time. The question whether the universe as a whole is in a superposition of states, as Hawking and Mlodinow (2010) have speculated, is a pseudoproblem. This is because nobody will ever be able to write a state equation for the universe, or perform an experiment to check that science-fiction conjecture. Writing down symbols purporting to designate state vectors for the universe is just hand-waving.

In sum, an electron is commonly in a superposition of elementary states. However, when it interacts with its environment, such superposition may contract to a minute cube or, ideally, to a point in space. This is what used to be called "collapse" (or "projection") of its state vector, the famous ψ. Nowadays this process is usually called *decoherence* (see, e.g., Schlosshauer 2007). Such a contraction of the probability cloud to a point may happen naturally, as when an electron goes through a very small pore. The same process also occurs when the position of the electron is measured with great precision. Something similar happens with the remaining dynamical properties of the quantons or quantum objects, whether small or large. However, Marcello Cini (1985) and others have argued persuasively that in these cases there is no instantaneous collapse but a more less quick contraction as a result of the quanton-macrothing interaction.

What matters philosophically is that the decoherence (or projection) process is purely physical, even when under experimental control. Observers may design or execute experiments, but their minds cannot act directly upon anything, because there is no such thing as telekinesis.

15.5 The Quantum Theory Is Not About Observers

A semantic analysis of the basic concepts of the quantum theory, such as the energy operator (hamiltonian) and the state vector or wave function, shows that they do not contain any variables referring to an observer. For instance, the energy operator for the simplest physical system, the hydrogen atom, consists of the kinetic energy plus the electrostatic energy of its electron. This suffices to calculate the possible energy levels of the atom in question, as well as the corresponding elementary state functions (eigenfunctions)—and much more besides.

The experimenter and the instruments she uses to excite the atom and measure the wavelength of the light that the atom emits when it decays from an excited energy level are in the laboratory, but they are not represented in the theory of the atom. Hence, the standard claim that the theory in question describes laboratory operations is utterly false. To describe such operations one must analyze the corresponding measuring instruments, all of which are macrophysical, with the help of classical physics, as well as indicators (see Chap. 8).

The result of such analysis is a rich panoply of special theories. There are as many such instrumental theories as types of instrument: the theory of the simple pendulum measuring the acceleration of gravity; the theory of galvanometer for measuring electric current intensities; the theory of the mass spectrometer for measuring atomic and molecular masses, and so on. Incidentally, the vast majority of these instrumental theories are applications of general classical theories, such as wave optics, which ignores photons. In conclusion, neither the instrumental theories nor the substantive quantum theories, such as those of the electron, refer to observers: all of them are strictly physical.

Note that this conclusion is not an arbitrary philosophical fantasy: it derives from an examination of the variables occurring in the theories in question. By contrast, the subjectivist thesis that the quantum theory is nothing but a description of the appearances experiences by an observer, is a groundless philosophical claim. This thesis is blatantly false if only because, as Galileo (1953) argued four centuries ago, physics knows nothing about secondary properties, such as colors, smells, tastes, and textures. All these properties, the *qualia*, emerge in brains, normally in response to physical stimuli characterized by primary properties, such as extension, duration, mass, and acidity. Besides, that thesis is a graft of the phenomenalism upheld by Berkeley, Hume, Kant, Mill, Comte, Mach, and the Vienna Circle.

If the phenomenalist (or positivist) interpretation of the quantum theory were correct, this theory would describe mental processes. Hence it would be part of psychology, not of physics, so that psychologists would not suffer from physics envy. Moreover, if that were the case, psychiatrists could use quantum physics to diagnose and treat mental disorders, such as the refusal to admit the reality of the external world.

15.6 Quantum Theory and Realism: An Entangled Affair

In their famous criticism of quantum mechanics, Einstein, Podolsky and Rosen (1935)—henceforth EPR—noted that this theory implies what was later called *entanglement*. This consists basically in the following fact: If two or more quantons initially constitute a system, they lose their individuality and seem to continue being united even after coming far apart.

For example, suppose that two quantons of the same class, such as electrons or photons, join forming a system with total energy E. Later on, either naturally or as an experimental effect, the two system components become spatially separate. In particular, they may fly off in opposite directions at birth, as is the case with the pair of photons emitted by an excited atom that decays to a lower energy state. The total energy does not change, but is now distributed at random among the two system components, in such a manner that neither of them has a sharp energy value. In other words, each component is in a superposition, or linear combination, of elementary energy states (eigenstates); and in turn the two state vectors are mutually entangled. Entanglement may then be seen as a special case of superposition. However, the converse view, of superposition as self-entanglement, is also tenable.

Our story does not end here but in disentanglement, just as superposition or coherence ends up in decoherence or projection. Indeed, sooner or later one of the system components, say the one labeled 1, interacts with a macrosystem that may, but need not, be a measuring apparatus. When this happens, the superposed state projects onto one of the elementary states, which corresponds to a sharp value E_1 of the energy.

That is, component 1 has now acquired a new property: not that of possessing energy, but that of having a precise amount of energy rather than a fuzzy one. And component 1 acquired this property because it interacted with a macrophysical thing, not because it was subjected to observation, since the process in question may be automated, so that it will also happen in the absence of any observers. And here comes the surprise: As the energy distribution of component 1 contracts to the value E_1, that of component 2 contracts to the precise m value $E_2 = E - E_1$.

The upshot is that the two quantons are now disentangled from one another: the system is no longer. Moreover, *disentanglement* occurs regardless of the distance that separates the two components, and in the absence of any signals between them. The maximal distance attained so far is 150 km, which separates observatories placed on two Canary islands.

EPR (1935) thought that entanglement constitutes "a spooky action at a distance". (Actually there may not be such *action*, since neither theory nor experiment involves signals or forces between the system's components.) And they believed that this and other counter-intuitive results of quantum mechanics defied any "reasonable definition of reality." This is why they judged that this theory was not realistic. And half a century later, when Alain Aspect proved the *reality* of entanglement, the great journal *Science* announced that *realism* had been refuted.

Throughout this crisis physicists tacitly used a reality criterion that, unlike EPR's, was not tied to classical mechanics. I submit that this alternative criterion is roughly the following. Let a and b name two different objects, and assume that a proves to be real at some time t. Then b too will be judged to be real at t, just in case either a acts upon b, or b acts upon a, at some time. (Note that this a criterion, not a definition. And note also that it should be refined in the light of the relativistic requirement of Lorentz invariance.)

Nowadays it is usually held that "local realism is untenable." This is sheer confusion. What is untenable is to call 'local realism' what is nothing other than the classical physical hypothesis that all changes are local or propagate by contact, so that the local can be uncoupled from the global. Whoever denies "local realism" should admit "global realism". However, the oxymoron "local realism" is best forgotten.

Allow me to insist. It is absurd to hold that "local realism is untenable," since what has been experimentally refuted is not *philosophical* realism but the *physical* hypothesis of separability, that the bond between *any* two things weakens as their mutual distance decreases until they behave independently from one another (Bunge 1989). If the things are quantons, once united they remain correlated regardless of their mutual distance.

If we were to give up genuine realism—that is, the principle that the physical objects exist outside the subject's mind—no experiments would be needed to find out what the world is like: it would suffice to ask the opinion of our favorite guru, or even to practice introspection.

In my opinion, both Einstein and the Copenhagen faithful were wrong concerning this matter. Einstein erred in believing that physical reality is classical, and Bohr in rejecting philosophical realism. The sensational successes of the quantum theory prove both that it is realistic, and that reality is not classical. In short, *Realism≠Classicism.*

That the quantum theory is counterintuitive or non-classical, is obvious but irrelevant. For example, this theory lacks classical position variables, that is, functions that assign an exact point in space to each quanton at any instant. Einstein believed that such a precise position and the corresponding precise trajectory are "elements of reality", so that their absence in quantum mechanics proved that this theory is incomplete.

This complaint motivated the construction of "hidden variables" theories, such as that of David Bohm, as well as Bell's theorem. (By definition, a hidden variable is one without "indeterminacy" or intrinsic variance.) John Bell proved in 1966 that every theory involving hidden variables must satisfy certain inequalities, namely Bell's, that are empirically testable.

A pile of experiments, like that of Alain Aspect in 1982, refuted Bell's inequality and thus the whole family of hidden-variables theories. And of course every Bell inequality violation is regarded as a confirmation of quantum mechanics. (In general, the refutation of *p* amounts to the corroboration of not-*p*. Hence, Popper notwithstanding, refutation is not stronger than confirmation.) By contrast, a vastly more numerous and varied set of experiments performed over nearly one century have shown that the quantum theory is the most precise of all physical theories.

Let us recall that every radically new theory may shock us and force us to reeducate our intuition. But if found true it is because it fits reality. Whoever gives up philosophical realism also gives up all hope of finding objective truths, as is the case of the constructive-relativist sociologists of knowledge. Science should not be confused with science fiction or with fantastic literature and art.

Anton Zeilinger (2010: 286), whose team has performed the most sensational measurements on entangled photons, stated that "[a]t present, there is no agreement in the scientific community as to what the philosophical consequences of the violation of Bell's inequality really are. And there is even less agreement about what position one has to assume now".

I submit that an agreement might easily be reached if EPR's idiosyncratic definitions of 'reality' and 'realism' were discarded, and a dictionary were consulted. An object other than the universe as a whole is real if any change in it causes a change in some other object. And philosophical realism is the doctrine according to which the universe is real. Let us not forget that the world existed before idealist philosophers.

Something similar applies to Stephen Hawking's skepticism about reality (Hawking and Mlodinow 2010). He claims that scientists judge an item X to be real if a model of X agrees with observation. Consequently, he claims that all we have is "model-dependent reality." But this conclusion holds only if reality is confused with truth, since an empirical confirmation of a conceptual model of X suggests that the said model is true, not that X is objectively real. For instance, a psychological model of a visual illusion may be true although the illusion, though real, is not objective since it happens inside the subject's head.

Moreover, a model may fit the data in spite of postulating an unreal entity, as was the case with classical wave optics, which involved the fictitious ether. Only fictions, like fairies and mathematical objects, are model-dependent: real things, by definition, exist independently of models.

In sum, the legitimate debate over the relation between quantum mechanics and reality has often been marred by *ad hoc* definitions of 'real' and 'realism'. Moreover, all of the discussants have ignored the elementary fact that, in evaluating the many perils that threatened his existence, primitive man could not wait for the emergence of philosophy, let alone of quantum mechanics: he had to be able to discover whether a rustling of leaves, or a carnivore's bad breadth, were reliable indicators of the reality of a nearby leopard.

15.7 Has Matter Vanished?

So far we have dealt with realism, the doctrine that the external world exists on its own and can be known. Let us now deal with materialism, the other philosophy often said to have been refuted by quantum physics. Materialism is the family of ontologies that hold the universe to be material. Most materialists admit the reality of the mental, but they claim that, far from being immaterial, every mental process is a brain process.

Materialism is the *bête noire* of all the religions and spiritualist doctrines and practices. Quantum idealism is only the fanciest of all the members of the contemporary immaterialist camp. It may have driven Wolfgang Pauli, the great quantum theorist and Copenhagen crusader, to flirting with the occult "sciences" and with Jung's archetypes. Homeopathy is a less presentable member of the same camp. Its founder, Samuel Hahnemann, claimed that a drug is the more effective, the less material. This is why he prescribed minimizing, through successive dilutions, the amount of active ingredient. However, let us go back to quantum physics.

The opinion that the quantum theory refutes materialism is suggested by its subjectivist interpretation jointly with the fact that the theory does not assign its referents the impenetrability of the bodies studied by classical mechanics. However, optics, born in antiquity, dealt with material things lacking in mass and solidity, namely light beams. The modern philosophers failed to take note of this tacit generalization of the notion of matter. Even the great Bertrand Russell believed that impenetrability was the defining trait of matter.

But what about antimatter: does it not challenge materialism? Not at all, because it is just a silly misnomer—like calling 'antianimals' the Antipodean animals. Indeed, the positrons, negative protons, and other species of "antimatter" have the same properties as their ordinary counterparts: mass, electric charge, spin, etc. Something similar holds for the "creation" and "annihilation" operators in quantum electrodynamics. In fact, the so-called creation of an electron-positron pair is only the transformation of a gamma photon into a electron and a positron; and the so-called annihilation is the converse transformation.

15.8 Indeterminism

Another popular belief is that quantum physics is indeterministic because its basic laws are probabilistic. This view is untenable even if Laplace's version of determinism is kept, for it fails for a collection of marbles. Indeed, Ludwig Boltzmann showed more than a century ago that in this case objective chance intervenes along with causation. And this calls for statistical mechanics, where the concepts of probability and randomness are central.

Both statistical and quantum physics suggest a broader and more subtle conception of determinism (Bunge 1959a). This conception boils down to the conjunction of the principles of lawfulness (causal, probabilistic or mixed) and of conservation of matter (though not of mass). Continuum mechanics, thermodynamics and quantum mechanics, which do not fit into Laplace's determinism, satisfy the updated version of determinism because they are centered in laws, and they contain conservation principles (or rather theorems).

The lawfulness principle denies that everything thinkable is really possible—Paul Feyerabend's infamous slogan *Anything goes*. For example, energy can be neither emitted nor asorbed in arbitrary amounts. And the conservation laws, whether classical or quantal, deny that matter may pop out of nothing. Besides, not all of the conservation laws are probabilistic. For example, the statement that the total angular momentum (orbital plus spin) of a particle in a central force field is constant, holds at every instant, not only on the average.

In summary, quantum physics did not kill determinism: it only prompted its renewal—which was overdue anyway.

15.9 Is Rationality Limited?

Finally, it has been claimed that the quantum theory sets limits on rationality, in that it is inherently obscure and proves that we cannot know all we would like to. Indeed, Bohr held that we must alter the meaning of the verb 'to understand', and Richard Feynman famously stated that "nobody understands quantum mechanics". However, neither Bohr nor Feynman told us explicitly what they meant by 'understanding'.

There is reason to believe, though, that what they had in mind is the ability to describe quantum facts in an intuitive or classical fashion. But it is not written that nature cares for our ability to intuit it. Classical physics is full of examples of processes that people had a hard time to understand, such as motion in the absence of forces, the diffraction of light, and the frame-dependence of length.

As for the limit on knowability, it is inherent in Heisenberg's original formulation of his celebrated "uncertainty principle" as the impossibility of knowing at the same time the precise position and the precise momentum of an electron. But this formulation is incorrect, because the premises that entail the said theorem do not involve the concept of knowledge: they are exclusively about quantons. What Heisenberg's theorem says is that electrons and their like do not *have* a precise position and a precise momentum at the same time. And it stands to reason that it is impossible to know something that does no exist. Thus, contrary to Heisenberg's assertion, the quantum theory is not about our knowledge of nature: it is a part of physics, not of epistemology or cognitive science.

In sum, the advent of the quantum theory was a victory of rationality and a huge addition to human knowledge.

15.10 Conclusion

To sum up, when the quantum theory is freed from its philosophical grafts, it is realized ha it is just as realistic, materialist, and deterministic as classical physics. What is true is that it forces us to update our ontology. In particular, it forces us to broaden the concepts of matter and determination, to downgrade the concepts of shape and individuality, and to reinforce the systemicity principle—due to the ubiquity of entanglement and decoherence.

References

Born, Max. 1953. Physical reality. *The Philosophical Quarterly* 3: 139–149.

Bunge, Mairo. 1959a. *Causality: The place of the causal principle in modern science*. Cambridge, MA: Harvard University Press.

Bunge, Mairo. 1959b. *Metascientific queries*. Springfield: Charles C. Thomas.

Bunge, Mairo. 1967. *Foundations of physics*. Berlin/Heidelberg/New York: Springer.

Bunge, Mairo. 1973. *Philosophy of physics*. Dordrecht: Reidel.

Bunge, Mairo. 1979. The Einstein-Bohr debate over quantum mechanics: Who was right about what? *Lecture Notes in Physics* 100: 204–219.

Bunge, Mario. 1985. *Treatise on basic philosophy*, vol. 7. Dordrecht: Reidel [Springer].

Bunge, Mairo. 1989. The bell inequalities and all that. *Philosophia Naturalis* 26: 121–134.

Bunge, Mairo. 2010. *Matter and mind*. London: Springer.

Cini, Marcello. 1985. Quantum theory of measurement without wave packet collapse. *Nuovo Cimento* 73B: 27–56.

Einstein, Albert, Boris Podolsky, and Nathan Rosen. 1935. Can quantum-mechanical description of physical reality be considered complete? *Physical Review* 47: 777–780.

Galilei, Galileo. 1953 [1623]. Il saggiatore. In *Opere*, ed. Flora Ferdinando. Milano-Napoli: Riccardo Ricciardi.

Hawking, Stephen, and Leonard Mlodinow. 2010. *The grand design*. New York: Bantam.

Heisenberg, Werner. 1958. *Physics and philosophy*. New York: Harper & Brothers.

Henry, Richard Conn. 2005. The mental universe. *Nature* 436: 29.

Mahner, Martin. 2001. *Scientific realism: selected essays by Mario Bunge*. Amherst: Prometheus Books.

Mermin, N.David. 1981. Quantum mysteries for anyone. *Journal of Philosophy* 78: 397–408.

Mosterín, Jesús. 2006. *La naturaleza Humana*. Madrid: Espasa Calpe.

Niels, Bohr. 1934. *Atomic theory and the description of nature*. Cambridge: Cambridge University Press.

Pauli, Wolfgang. 1961. *Aufsätze und vorträge über physik und erkenntnistheorie*. Braunschweig: Vieweg & Sohn.

Roland, Omnès. 1999 [1994]. *Quantum philosophy*. Princeton: Princeton University Press.

Schlosshauer, Maximillian. 2007. *Decoherence and the quantum-to-classical transition*. Berlin/Heidelberg/New York: Springer.

Strawson, Galen. 2008. *Real materialism and other essays*. New York: Oxford University Press.

Zeilinger, Anton. 2010. *Dance of the photons: From Einstein to quantum teleportation*. New York: Farrar, Straus & Giroux.

Chapter 16
Parallel Universes? Digital Physics?

Every original idea is imaginative, because only imagination can trigger creativity. This is why imagination is just as essential in science and technology as in the arts and humanities. The difference between these two pairs of fields is that in science and technology imagination is disciplined rather than free. What motivates such discipline is the objective truth requirement.

In this chapter we will examine the multiverse fantasy. I submit that this is a case of methodological naiveté, on top of which it does not solve any cosmological problem, such as that of the nature of the so-called dark matter. We shall also examine digital physics, according to which the bricks of the universe are not elementary particles but bits, or information units. We shall argue that this opinion is but the latest attempt to dematerialize the world for the sole benefit of idealist metaphysics and theology (see e.g. McMullin 2010*).*

16.1 Imagination: Free and Disciplined

The discipline reigning in the scientific community is double: logical and empirical. Indeed, it is normally required that the original idea be (a) coherent and expressed in clear language; and (b) empirically testable at least in principle. Most scientists fulfill both conditions. But sometimes they fail to do so, in the manner of drivers that occasionally break transit rules. Fortunately their peers take pleasure in subjecting them to critical scrutiny, so that harm is usually contained.

In science there are several sources of unbridled speculation. Among them we find mathematical imperialism, methodological naiveté, and philosophical idealism, along with its big sister, theology. I call *mathematical imperialism* the doctrine according to which mathematics suffices to investigate real things. The clearest and most dangerous example is the bulk of mathematical economics, for it ignores empirical data (recall Chap. 8).

Methodological naiveté, or anarchism, as Feyerabend (1978) called it, is the thesis that "anything goes": that scientists should not bother about the testability of their

M. Bunge, *Evaluating Philosophies*, Boston Studies in the Philosophy of Science 295, DOI 10.1007/978-94-007-4408-0_16,

ideas. This doctrine is the philosophical equivalent of the sale of pharmaceutical products that have been neither analyzed nor tested. Such naiveté is obvious in current evolutionary psychology, which abounds in imaginative explanations of social facts, real or imaginary, that are either untestable or incompatible with findings in the social sciences. A popular instance of such naiveté is the thesis that we are walking fossils, since our mind was shaped once and for all when our remote ancestors had to cope with predators in the Africa savanna (see Buss 2004). Never mind that the sensational advancement of knowledge refutes the fantasy.

In sum, science is bounded poetry—just as far from doggerel as from surrealism.

16.2 Possible Worlds

Starting with the "discovery" of America there have been philosophers, political thinkers and novelists who imagined physical or social worlds different from ours. Ironically, whereas some theists have held that ours is the best of possible worlds, some atheists claim that the existence of multiple universes refutes the fantasy of intelligent design. However, it is obvious that an omnipotent Being, who can create things just by naming them, can create many worlds just as easily as a single one.

Whereas some multiverse theorists regard the alternative universes as fictions, others believe that they are just as real as ours. Here we shall deal exclusively with the latter and, in particular, with the parallel physical universes that some cosmologists and physicists have imagined in the course of the last half a century.

Some distinguished scientists believe that there are universes parallel to ours and disconnected from it. (Do not ask how they know it.) This belief, usually called the *multiverse* hypothesis, has been taken seriously by some cosmologists and string theorists (e.g., Tegmark 2003). Some of the latter estimate that there are no less than 10^{500} universes—an unthinkable number.

This idea is all the more remarkable for having been proposed by people who have not succeded in explaining why they have invested such huge ingenuity, over almost half a century, in attempting to justify the extravagant conjecture, essential to string theory, that our universe has ten spatial dimensions rather than three.

Another source of the possible worlds fantasy is the interpretation of quantum mechanics proposed by Hugh Everett III (1957) in his doctoral dissertation supervised by John A. Wheeler. Let us peek at it. Suppose that, as is usually the case, the energy state (or state vector) of a quantum-mechanical object, such as an atom, is blunt or smeared rather than sharp. More precisely, let the state in question be a linear combination of elementary energy states, each with its weight or probability. Such a state is said to be *coherent*.

So far our atom has been isolated. If now its energy is measured in an invasive fashion, decoherence occurs, and a sharp energy value is obtained (see Chap. 10). What happens to the other possible values? Common sense would reply: they vanish, just like any other potentiality that has not been actualized, such as the tail when a coin lands head up. Everett suggested, instead, that every one of the unrealized possibilities in our world is actualized in an alternative universe, which would be just as real as the one that hosts us.

This extravagant reply does not tell us where the energy of the alternative universes would come from, nor what evidence there is for their existence, nor for the conjecture that every one of them contains a counterpart of our favorite experimenter. We are only told that the alternative universes cannot communicate with one another. This additional hypothesis guarantees that the alternative universes are unobservable, so that the hypothesis is irrefutable. Still, Everett's parallel universes are less crazy than those invented by some cosmologists, for at least the former have the same laws and the same values of the fundamental constant as ours.

What is the scientific status of the parallel universes fantasies? Is it guaranteed by the fact that they have been published in first rate scientific journals such as *Physical Review* and *Astrophysical Journal*? This is what a social constructivist might claim. For the rest of us that fact only shows that not all the referees of scientific journals are endowed with pseudoscience detectors—or with the courage required to recommend the rejection of submissions by great names backed by prestigious academic bodies.

If we admit that testability in principle and compatibility with the bulk of scientific knowledge are peculiar to scientific hypotheses, then the many-worlds hypothesis must be rejected out of hand, because the alternative worlds are assumed to be inaccessible from one another, in particular from ours. Brian Greene (2011:165), an enthusiastic believer in multiverses of various kinds, rejects this objection and adopts what he calls the "expansive viewpoint", that tolerates untestable hypotheses about occult domains, hoping that they may eventually be subjected to test. Fortunately there still are scientists intolerant of unbridled speculation.

Note that we have demanded the *testability*, not the *refutability*, of hypotheses. The reason is that in the sciences the existential hypotheses call for positive evidence. In fact, whoever holds that X exists has the duty to exhibit data showing that X exists or can exist in the real world. Think of the hypotheses of the existence of Uranus, radio waves, transuranian elements, genes, HIV, stress, antipodeans, preadamites, or matriarchal societies.

In summary, the multiverse hypotheses are escapist and pseudoscientific. Moreover, they are barren, since it is impossible to study those extra universes.

16.3 Virtual Worlds? Virtual Particles?

The multiverse fantasies have nothing to do with the "virtual worlds" of painters, engineers, ad men, and computer buffs. These "virtual worlds" do not exist outside the screens and the brains that perceive them: without the eye, or rather the brain, there is no deception either.

Yet, the virtual realities are just as real as those brains, since these conjure them up, just as they dream and hallucinate. The presence of something in the world external to a subject is sufficient but not necessary for its real existence: their internal world is in my external world. Let us now look at a harder problem.

What is the status of the virtual particles occurring in theoretical particle physics: are they real or fictitious? This is the reply of the Stanford Linear Accelerator people (SLA 2001): "Virtual particles are a language invented by physicists in order to talk about processes in terms of the Feynman diagrams. These diagrams are a shorthand for a calculation that gives the probability of the process." Shorter: Virtual particles are only computational devices.

Consequently it is foreseeable that people will cease to talk about virtual particles as soon as the Feynman diagrams are superseded. Meanwhile there will be methodologically naïve physicists who will tell us that the virtual particles are real but unobservable because they have very short lives, during which they have the luxury of violating Einstein's fundamental equation that relates their energy to their momentum and mass,

To repeat, the virtual particles are fictitious and therefore unobservable. They are but artifacts invented to use Feynman diagrams. These, in turn, are heuristic tools that facilitate complicated calculations in quantum electrodynamics. Nature does not "use" heuristic artifacts. (More in Bunge 1959, 1970.)

16.4 Digital Physics

The imaginative American physicist John Archibald Wheeler tried, over more than half a century, to expel matter from the basic science of matter. His first attempt to subordinate physics to idealist metaphysics was geometrodynamics (Wheeler 1962). This was a mathematically sophisticated theory that became famous for its wormholes, and which claimed to construct the world out of space and time. But this theory failed, as it did not solve any extant problems, and its debris proved useless to construct further theories.

Wheeler's last attempt to free physics from the concept of matter was digital physics (see Barrow et al. 2004). The central idea of this theory is that the bricks of the universe are bits, the information units. Wheeler compressed it into the formula *Its from bits*. This is a new version of the Pythagorean myth, as Gregory Chaitin (2006), a prominent member of the new esoteric sect, admits.

To see how the things-from-symbols magic works, consider the process of getting and drinking a glass of milk, every step of which we assign a string of 0s and 1s as follows:

Open fridge	00
Take out glass of milk	01
Drink milk	10
Close fridge	11

The whole process can thus be encoded or represented by this string: 00011011. It would be lunacy to think that this symbol *equals* that process, but this is precisely what the its-from-bits thesis says.

This idea seems to have had several sources. One of them, obvious in the case of Wilczeck (2008) and other distinguished theorists, is the confusion of real things with our symbols or models of them. This confusion is a trait of the magical thinking involved in the practice of inserting pins in dolls as a sure and low-cost assassination procedure. Pythagoras or one of his brothers may have reasoned thus: Since everything that surrounds us comes in precise quantities, numbers are in everything, and perhaps all things are numbers.

This jump from things to models, or from ontology to semiotics and epistemology, is typical of archaic thought but not exclusive to it. For example, Vlatko Vendral (2010:10), Oxford professor of Quantum Information Science, writes: "information (and not matter or energy) is the building block on which everything is constructed. Information is far more fundamental than matter or energy because it can be successfully applied to both macroscopic interactions, such as economic and social phenomena, and, as I will argue, information can also be used to explain the origin and behaviour of microscopic interactions such as energy and matter."

Recall the fallacy of our Pythagorean friend: The world is made out of numbers because these can be used to describe everything. Furthermore, accounting the lower by the higher, and not conversely as it is usual in science, confer the ideological respectability and the social prestige that comes with it. But the idea that nature's building blocks are artificial is self-contradictory. The same holds for the claim that evolution is guided by algorithms (Dennett 1995), whence natural selection would be artificial.

16.5 Physics ≠ Semiotics

It is time to ask whether there is any truth to digitalism. A first obvious objection is that artifacts are preceded by natural objects, both historically and ontologically, not the other way round. In particular, contrary to what John asserted in his Gospel, the word came only recently, not in the beginning. In general, symbols are made, not found. And the earliest known symbols belong in cave paintings, which were made only about 30,000 years ago.

Secondly, whereas the natural things are essentially active, signs are inert: they cannot do anything by themselves. In particular, symbols can convey information only to non-symbolic things such as animals and machines. (Do not confuse signs with signals: these are processes.) Hence, a world made of symbols would be static. So much so, that no one has dared propose laws of motion of symbols—and without such laws there is no physics.

Thirdly and as a consequence, semiotics, the science of signs, cannot replace the natural sciences, just as proof readers are not qualified to repair hernias.

In the fourth place, there is no known mechanism whereby symbols generate natural things. Symbols are characterized by perceptible traits such as shape, not by imperceptible ones such as energy. How could signs, such as 0 and 1, which lack energy and obey man-made rules rather than laws of nature, generate bits of matter such as electrons and photons? Does not the formula *its from bits* evoke the Catholic ceremony of transubstantiation, whereby the priest transforms a wafer into Christ's body and wine into his blood? It is not by accident that it has been claimed that the information concept bridges science and Christian theology (Davies y Gregersen 2010).

Fifth, the formula in question cannot acquire any empirical support, because signs depend on conventional codes and are devoid of primary properties, whence they cannot be subjected to laboratory operations. One cannot even conceive of a mental experiment involving a real process capable of transforming a string of symbols into a material thing—short of invoking paranormal faculties, the way the Bible does when narrating God's exploits during that memorable week, when all He did was to utter orders of the form "Let X be!".

The sixth and last objection is that, although an information *theory* may *pretend* that information exists without a material support, real information systems, such as computers and television networks, are material systems designed by engineers and assembled in factories. In sum, what is true is the formula *bits from its*, not its converse.

In short, digital physics is very remote from both physics and information technology.

16.6 Concluding Remarks

The multiverse doctrines and digital physics are fantasies in scientific disguise. They are incompatible with the bulk of science and they do not solve any pending scientific problems. They only raise a psychological problem: How could so many distinguished scientists have indulged in pseudoscience?

Such combination of solid scientific achievement with pseudoscience is not new. Charles Richet, who earned the Nobel Prize for discovering allergy, presided over the French parapsychology society; René Blondlot, a distinguished physicist, claimed to have discovered the N rays, that proved to be imaginary; Sir Arthur Eddington, the eminent astrophysicist, believed that he had deduced the universal physical constants (such as c and G) from a priori principles; J. B. S. Haldane, the great geneticist and cofounder of the synthetic theory of evolution, defended the vernalization fantasy proposed by Lysenko; Sir John Eccles, a Nobel Prize in neuroscience, believed that the immaterial mind activates the brain; and Luc Montagnier, who got the Nobel Prize for having discovered the AIDS virus, is defending homeopathy. Obviously, a rich scientific experience does not immunize against pseudoscience: a dose of pro-science philosophy too is needed.

References

Barrow, John D., W. Paul, Davies, L. Charles, and J.R. Harper (eds.). 2004. *Science and ultimate reality: quantum theory, cosmology, and complexity*. Cambridge: Cambridge University Press.

Bunge, Mario. 1959. *Metascientific queries*. Springfield: Charles C. Thomas, Publisher.

Bunge, Mario. 1970. Virtual processes and virtual particles: Real or fictitious? *International Journal of Theoretical Physics* 3: 507–508.
Bunge, Mario. 1983. Speculation: Wild and sound. *New Ideas in Psychology* 1: 3–6.
Bunge, Mario. 1998. *Social science under debate*. Toronto: University of Toronto Press.
Buss, David M. 2004. *Evolutionary psychology: The New science of the mind*, 2nd ed. Boston: Pearson Education.
Chaitin, Gregory. 2006. *Meta math!* New York: Viking.
Davies, Paul, and Niels H. Gregersen (eds.). 2010. *Information and the nature of reality*. Cambridge: Cambridge University Press.
Dennett, Daniel C. 1995. *Darwin's dangerous idea*. New York: Simon & Schuster.
Everett III, Hugh. 1957. "Relative state" formulation of quantum mechanics. *Reviews of Modern Physics* 29: 454–462.
Feyerabend, Paul K. 1978. *Against method*. Repr. London: Verso.
Greene, Brian. 2011. *The hidden reality*. New York: Knopf.
Hawking, Stephen, and Leonard Mlodinow. 2010. *The grand design*. New York: Bantam.
McMullin, Ernan. 2010. From matter to materialism—And (almost) back. In Davies and Gregersen, eds., pp. 13–37.
SLAC. 2011. http://www.2.slac.stanford.edu//vvc/theory/virtual.html.
Tegmark, Max. May 2003. Parallel universes. *Scientific American* 288(5): 41–51.
Vendral, Vlakto. 2010. *Decoding reality: The universe as quantum information*. Oxford: Oxford University Press.
Wheeler, John A. 1962. Empty space-time as the building material of the physical world. In *Logic, methodology and philosophy of science: Proceedings of the 1960 international congress*, ed. E. Nagel, P. Suppes, and A. Tarski, 361–374. Stanford: Stanford University Press.
Wilczeck, Frank. 2008. *The lightness of being: Mass, ether, and the unification of forces*. New York: Basic Books.

Chapter 17
Can Functionalist Psychology Explain Anything?

Every science starts by describing facts in some sector of reality. And a science matures to the extent that it explains what it has described.

For example, Newton explained by gravitation and inertia the planetary orbits that Kepler had described; Spallanzani explained the "facts of life" by the union of an egg with a spermatozoid; and Weber explained the decadence of the so-called ancient world by the shrinking of the slave market, which in turn was due to the cessation of Roman territorial expansion.

All of the above explanations involve only scrutable material factors. This, not logical form, is the peculiarity of scientific explanation in contradistinction to explanations in supernatural, spiritual, or paranormal terms: that it is compatible with the materialist (though not necessarily physicalist) ontologies. Is this the rule in psychology too? Let us see.

17.1 Functionalist Description

Since its birth psychology has been accounting for mental facts in various ways. The simplest explanation of such facts involves only other mental facts, as when we say that she did that because she believed that it was in her own interest to proceed thus. But such description does not explain anything, because it does not involve any mechanism transforming belief into action. Indeed, to explain a fact is to unveil the underlying mechanism (Bunge 2006).

How can one go from description to explanation in psychology? Such a transition can be accomplished by replacing the immaterial mind with the brain, because this is the organ of the mind, and because mechanisms exist only in material systems. Let us begin with an analogy.

The instructor of the driving school describes the functions of the parts the student is expected to manipulate. For example, he will tell her "If you press this pedal, the car will accelerate." He will not necessarily tell her that the pedal regulates the flow of gasoline from the tank to the cylinders, where it explodes due to the spark

M. Bunge, *Evaluating Philosophies*, Boston Studies in the Philosophy of Science 295,
DOI 10.1007/978-94-007-4408-0_17, © Springer Science+Business Media Dordrecht 2012

produced by the spark-plugs; that the explosion pushes the piston, and so on. The student is not interested in the underlying mechanisms: she is only interested in functions. But of course mechanisms are of paramount interest to the engineers who design car, as well as to the mechanics who maintain or repair them: they must know how things work.

In a similar fashion, functionalist psychology handles mental functions (or processes), such as remembering and imagining, independently of their material "substrates". For example, the functionalist may explain a placebo effect as caused by a belief, or occasional stuttering as the effect of a strong emotion. A classical representation of such explanations is by diagrams consisting of boxes (functions) joined by arrows (causal connections or just successions). For example,

Perception → *Short term memory* → *Long term memory*,
and
Deliberation → *Decision* → *Action*.

Assuming that such diagrams are faithful representations of true propositions, are they enough? No, because they are disembodied: They overlook the definition of 'process' as change in a material thing.

In particular, mental events are changes in the brain, not in the immaterial soul or mind imagined by shamans, theologians, and philosophers.

Imagine a functionalist physicist. He may tell, us for instance, that air warms up when the Sun rises; that heavy things fall down when released; and that thunder follows lightning. But he will not tell us why such associations happen. Functionalist physics is just folk physics. It fits the phenomenalist ontology of Hume, Kant, Comte, Mach, Carnap, Goodman, and David Lewis.

Scientific physics, by contrast, looks behind appearances: it explains the warming of the air by sunshine as an effect of the absorption of light by air molecules; the fall of bodies by their interaction with the Earth via their gravitational fields; and thunder by the sound shock waves generated by electrical discharges between clouds.

17.2 To Explain Is to Unveil Mechanisms

Whereas ordinary knowledge is confined to describing phenomena (appearances), science explains them in terms of more or less occult but scrutable mechanisms. Hence, if scientists had listened to the phenomenalists, they would not have invented modern physics. In particular, they would have invented neither field physics nor atomic physics, since phenomenalists deny the existence of fields and atoms for being unobservable.

Something similar happened in biology, psychology, and the social sciences: they have been explaining through mechanisms what folk wisdom takes for granted and much more. Here is a handful of examples.

1. Genetics explains that we resemble our ancestors because we have inherited some of their genes, in particular those that regulate morphogenesis.

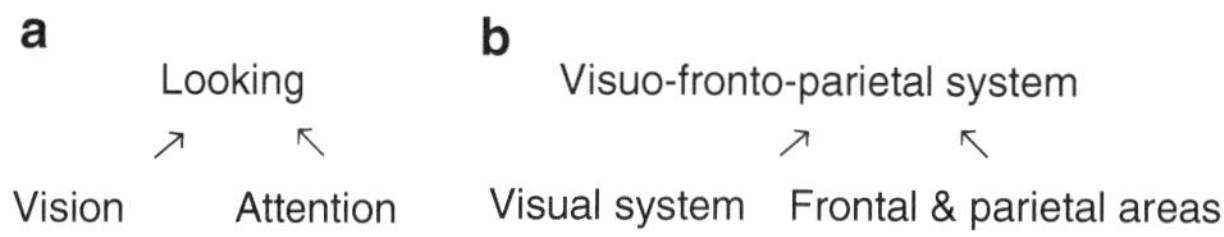

Fig. 17.1 Seeing/looking. (**a**) Functional description. (**b**) Neural explanation (Bressler et al. 2008)

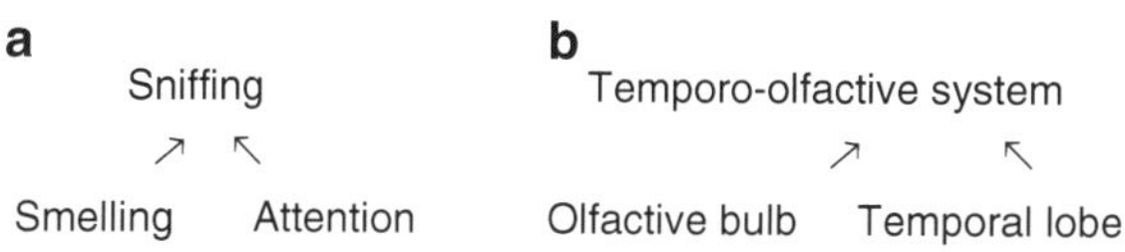

Fig. 17.2 Smelling/sniffing. (**a**) Functional description. (**b**) Neural explanation (Sobel et al. 1998)

2. This is how a mosquito finds us: A specialized neuron in its “nose” detects some of the CO_2 that we exhale, and generates a signal that is transmitted through the insects’ brain to its “wings”, leading it upwind to its meal—us.
3. Memories are neuronal configurations. These are initially unstable, but they get consolidated when the brain synthesizes certain proteins that glue the inter-neuronal connections—which is why repetition yields memorization.
4. The mammalian’s hippocampus “stores” memories, particularly spatial ones—which is why the London taxi drivers have larger-than-average posterior hippocampi.
5. At the molecular level, depression and other mental disorders consist in imbalances of neurotransmitters—which is why they can be treated with medications that alter neurotransmitter synthesis, release, break-down, or reuptake.
6. Physical exercise is known go improve mood and cognition, because it intensifies blood circulation, which in turn feeds the brain.
7. Looking differs from seeing, and sniffing from smelling, because the corresponding mechanisms are different, as shown schematically in Figs. 17.1 and 17.2.

Another example: literacy alters the cortical networks for language and vision (Dehaene et al. 2010). That learning to read facilitates speech comprehension is not surprising, particularly since it is well known that both functions are localized mainly in the left cerebral hemisphere. What is surprising is the finding that literacy decreases the ability to recognize faces. Whereas a functionalist would confine himself to recording this fact, Dehaene and coworkers (op. cit.: 1359) explain it tentatively as follows: Since writing is a comparatively recent acquisition, “the reading processes must invade and ‘recycle’ cortical space devoted to evolutionary older functions”.

Something similar might happen with the capacity for reasoning. Vygotsky and Luria had found, almost a century ago, that this capacity is strengthtened notably after just 1 year of elementary schooling, devoted almost exclusively to learning to read and write. The acquisition of these new skills would indirectly facilitate the reasoning ability, by changing connections in the neural networks.

That the brain changes as it learns, has been shown numerous times in animal research (e.g., Buonomano and Merzenich 1998; Scholz et al. 2009), and also in recent years through brain imaging in humans (e.g., Klingberg 2010). The same technique has also revealed the neural effects of urban upbringing and habitation, which can involve chronic stress (Lederbogen et al. 2011). True, the underlying mechanisms, in this case as in the previous ones, are still to be discovered. But at least it is now known where to look: the amygdala, the perigenual anterior cingulate cortex, and associated brain regions.

Lastly, a speculative conjecture: There are at least seven kinds of disgust—one for every sense organ plus intellectual and moral disgusts. Every one of them is the specific function of a distinct neural circuit, but all seven have a common intersection that contains the insula.

Since explaining a fact is to unveil its underlying mechanisms, the explanation of mental facts is in charge of physiological psychology, also called cognitive and affective neuroscience. However, this does not involve micro-reduction, since neural processes are influenced by social stimuli. Let us peek at this.

17.3 Submission Sickens

A famous example of the influence of the social environment upon the human body is the Whitehall immunological study, that investigated the physical and mental health of the British public servants (Marmot et al. 1991). This study showed that the morbidity rate rises as the runks in the hierarchy descend. The mechanism is this: Dependency increases stress, which stimulates the synthesis of corticoids, which harm the internal organs. (Incidentally, this is a good argument for responsible liberty and self-government.)

The functionalist skips the intermediate level and summarizes the Whitehall findings into this schema: ↑Dependency → ↑Morbility. But this is just a datum, albeit an important one. Hume and his followers would have been satisfied with it. The neuroscientists, by contrast, asked why such a strange sequence should occur: they sought the underlying mechanism, and found part of it, namely this:

↑Dependency → Stress → ↑Corticoids → ↓Immunity → ↑Morbility.

What is more, the psychoneuroendocrinoinmunologists in question discovered new problems and thereby confirmed the old principle of scientific realism: Scientific research is endless because it is self-propelled. Schematically,

Problem → Solution → New problem → New finding →…

17.4 Beyond Nativism and Empiricism

Let us now jump from the organs level to that of cells. Donald Hebb (1949), arguably the founder of contemporary cognitive neuroscience, conjectured that learning consists in the formation of new cell assemblies, a process that may be triggered by internal or external stimuli.

Let us figure out how an infant may learn to see a triangle (Scott 2002: 260). We may conjecture that the baby perceives three different vertices in a row, and ends up by putting them together into a triangle. Translated in terms of Hebb's cell assemblies, this sequence reads thus: assembly 1 (for vertex 1) in the primary visual area is "ignited" by an external stimulus; then the same assembly 1 ignites assembly 2 (for vertex 2), which ignites assembly 3 (for vertex 3), which completes the triangle.

As with perception, so with conception. Thus a simple thought, such as "The baby cried", would be the same as the activation of a cell assembly somewhere in the cortex. And a train of thought, such as "The baby cried, then its mother picked it up," would be the sequential activation of a number of adjoining cell assemblies. Another intriguing hypothesis of Hebb's concerns the abstraction scale: The more abstract an idea, the farther away from the sensory cortex it would be located.

The neurobiological approach to mind allows us to evaluate Kant's famous hypothesis that space and time, far from being in the world, are in the mind and, more precisely, are "*a priori* forms of intuition." This conjecture contradicts a basic if tacit assumption of physics, as well as the widespread idea that everything real is or happens in space and time. It also contradicts an old finding of developmental psychology: that one learns to perceive, in particular shapes and processes.

The downfall of Kant's psychology does not vindicate the empiricist hypothesis, that at birth the mind is a *tabula rasa*, a blank tablet on which experience writes. It is not, because we inherit organs specialized in perceiving spatial and temporal traits. For example, it has recently been discovered that rat pups have, at birth, fully developed directional neurons, as well as place cells that increase in number and change considerably with age (Palmer and Lynch 2010).

Nothing of this proves that Kant and the other nativists, from Socrates to Leibniz to Chomsky, were right: it just refutes empiricist psychology, from Aristotle to Locke to Skinner. And it confirms the hypothesis that, although we are not born with ideas, we are born with the ideation organ. And this organ is, of course, a product of both evolution and individual development, which in turn is strongly influenced by the social environment.

Now, it so happens that functionalist psychology ignores both brain development and social milieu: it regards all persons as adults who compute whatever they feel, desire, think, and do, in accordance with innate and inflexible algorithms (computation rules). But we all know that we "think to rule", rather than in a zigzag fashion, only when calculating. And one always calculates following rules or algorithms, so that no new ideas can be expected to emerge in such process.

In other words, all computation is algorithmic, and nothing algorithmic is creative. By contrast, the human brain is plastic, to the point that having a learning and creative mind may amount to having a plastic brain (Bunge 1980).

Moreover, because they treat all minds as if they were adult, the functionalists cannot explain the changes in behavior that happen as we grow up. For instance, why do adolescents take more risks that either children or adults? Presumably, because during that intermediate phase the dopamine level in the brain rises significantly, at the same time that the self-control networks involving prefrontal cortex are still under-developed, while social controls relax, so that new opportunities emerge but no new responsibilities constrain. All this explains why the crime rate peaks at about 17 years of age and decreases thereafter. It also explains why criminologists, judges and law makers are paying increasing attention to cognitive and affective neuroscience (e.g., Wikström and Treiber 2007).

17.5 Concluding Remarks

In conclusion, functionalists can *describe* some mental facts but they cannot *explain* any in terms of processes, in particular mechanisms, because they have adopted an obsolete philosophy of mind, namely psychoneural dualism. Moreover, functionalists disregard everything that is not mental, so that they miss the strong interactions between the nervous, endocrine, and immune systems. Thus the functionalist approach seriously impoverishes the psychological problematics. Contemporary psychology, by contrast, is a mixed science: it is—in a word—*psychoneuroendocrinoimmunosociology*.

This synthesis was triggered by the philosophical hypothesis that mental facts are brain processes. In turn, the scientific investigation of the mental in the living brain led to the discovery that the brain is a part of a more comprehensive biological system, which in turn is closely linked with the social system. And this is a striking confirmation of yet another philosophical hypothesis, namely systemism. This blend of materialism with systemism deserves a name of its own: *systemic materialism*.

The practical dividend of this process of materialization of the soul is the recent transformation of psychiatry, from the most backward and inefficient branch of medicine, into an increasingly rigorous and efficient one. This change is due to the fact that the biological psychiatrists have adopted a systemic and materialist approach, and have started to use certain drugs that alter some of the chemical reactions occurring in the sick brain. Although the results of this recent change in approach are still rather modest, they have sufficed to discredit psychoanalytic shamanism as well as antipsychiatry (see Shorter 1997).

Regrettably, the vast majority of philosophers of mind have not helped. In exalting functionalism, in particular computationalism, and in disparaging the biosocial approach to the mind, they have slowed down the scientific understanding of the mental, thus indirectly harming the mentally ill as well as juvenile delinquents. Believers in Hell might say that the dualist philosophers have been unwitting helpers of the Devil, ever ready to snatch stray souls.

References

Bressler, Steven L., Wei Tang, Chad M. Sylvester, Gordon L. Shulman, and Maurizio Corbetta. 2008. Top-down control of human visual cortex by frontal and parietal cortex in anticipatory visual spatial attention. *Journal of Neuroscience* 28: 10056–10061.

Bunge, Mario. 1980. *The mind-body problem*. Oxford: Pergamon.

Bunge, Mario. 2006. *Chasing reality*. Toronto: University of Toronto Press.

Buonomano, Dean V., and Michael M. Merzenich. 1998. Cortical plasticity: From synapses to maps. *Annual Review of Neuroscience* 21: 149–186.

Dehaene, Stainslas, et al. 2010. How learning to read changes the cortical networks for vision and language. *Science* 330: 1359–1364.

Hebb, Donald O. 1949. *The organization of behavior*. New York: Wiley.

Hubel, David H., and Torsten N. Wiesel. 1962. Receptive fields, binocular interaction, and functional architecture in the cat's visual cortex. *The Journal of Physiology* 160: 106–154.

Klingberg, Torkel. 2010. Training and plasticity of working memory. *Trends in Cognitive Science* 14: 317–324.

Lederbogen, Florian, et al. 2011. City living and urban upbringing affect neural stress processing in humans. *Nature* 471: 498–501.

Mahner, Martin, and Mario. Bunge. 2000. Function and functionalism: A synthetic perspective. *Philosophy of Science* 68: 75–94.

Marmot, Michael G., et al. 1991. Health inequalities among British civil servants: The Whitehall II study. *The Lancet* 337: 1387–1393.

Palmer, Linda, and Y. Gary. Lynch. 2010. A Kantian view of space. *Science* 328: 1487–1488.

Scholz, Jan, et al. 2009. Training induces changes in white-matter architecture. *Nature Neuroscience* 12: 1370–1371.

Scott, Alwyn. 2002. *Neuroscience: a mathematical primer*. New York: Springer.

Shorter, Edward. 1997. *A history of psychiatry*. New York: Wiley.

Sobel, N., V. Prabhakaran, E.E. Desmond, G.H. Glover, R.L. Goode, E.V. Sullivan, and J.D. Gabrieli. 1998. Sniffing and smelling: Different subsystems in the human olfactory cortex. *Nature* 392: 282–288.

Wikström, P-Olov, and Kyle Treiber. 2007. The role of self-control in crime causation. *European Journal of Criminology* 4: 237–264.

Chapter 18
Knowledge Pyramids or Rosettes?

One says that biology is based on chemistry, which in turn is based on physics. This statement sounds plausible to most of us, but admittedly the expression 'is based on' is not clear. Let us try to elucidate it.

The most obvious idea is that some bodies of knowledge are ordered in a pyramidal fashion. (Notice the prefix 'some': there is more than one such pyramid.) But what exactly, is the precedence relation that occurs in the pyramid metaphor? This question presupposes that there is a single precedence relation, and logicians are likely to agree. But this is not so: the expresión 'is based on' is ambiguous, that is, it designates several concepts. (Nominalists, such as Tarski and Quine, disagree: they disown concepts in favor of terms.) More precisely, we are dealing with a whole family of concepts with the same logical form, namely <, but different senses or contents. (See Bunge 1974 *for clarifications of the concept of sense.)*

Let us explore some of those concepts: those of epistemological, logical, and ontological precedente. And let us suppose that we have a reasonably clear concept of body or system of knowledge, which may be formless like a research field or well structured like a theory.

18.1 Epistemological Pyramids and Rosettes

We say that the body of knowledge B is *epistemolgically* based on the body A, or $A <_E B$ for short, if knowledge of A is necessary for that of B. For example, Mathematics $<_E$ Physics. The relation $<_E$ may be said to occur with different intensities. We distinguish strong from weak precedente: strong when knowledge of the totality of A is necessary for that of B, and weak when only a part of A need be known to know the body B in question or to do research in it. Clearly, the weak precedence relation is vague, and yet the one that applies in practice.

The $<_E$ relation induces pyramids like that of Fig. 18.1:

This pyramid is not quite correct because it leaves no room to the biosocial sciences, such as anthropology and psychology, demography, and epidemiology.

M. Bunge, *Evaluating Philosophies*, Boston Studies in the Philosophy of Science 295,
DOI 10.1007/978-94-007-4408-0_18, © Springer Science+Business Media Dordrecht 2012

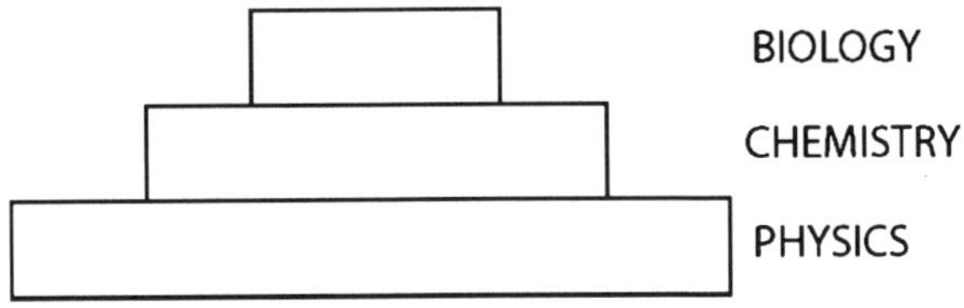

Fig. 18.1 Epistemological pyramid of the basic natural sciences

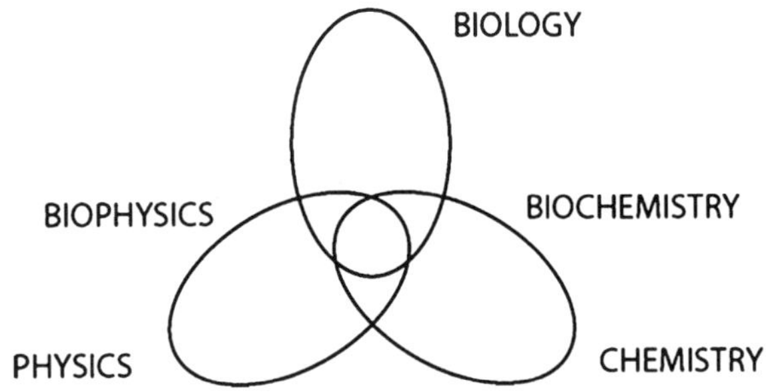

Fig. 18.2 Rosette of the natural sciences

Indeed, a hybrid discipline has more than one "basis" on the same footing. (See Bunge 2003 for the concept of an interscience.) For example, social cognitive neuroscience is based on cognitive neuroscience and sociology.

The disciplines of this kind are better represented by Venn diagrams (partially intersecting ellipses) than by pyramids, See Fig. 18.2:

18.2 Logical Pyramids

We may say that the body of knowledge A precedes *logically* the body B, or $A <_L B$ for short, when A contains some of the propositions that entail B. The relation of logical precedence may be said to be *strong* when A suffices to deduce B, and *weak* when A must be enriched with another body C for B to follow.

For example, dynamics strongly precedes both statics and kinematics; likewise, electrodynamics strongly precedes optics. (The precedente in question is logical, not historical.) On the other hand, contrary to a widespread belief, classical mechanics is no enough to deduce statistical mechanics: one must add the hypothesis that the initial positions and velocities of the molecules are distributed at random.

The weak precedente relation induces pyramids such as the following:

INFINITESIMAL CALCULUS	BIOLOGY
SET THEORY	CHEMISTRY
PREDICATE LOGIC	PHYSICS

The right hand side pyramid makes no room for biophysics or for biochemistry, since both result from the fusion of two disciplines on an equal footing. By contrast, all five research fields can easily be represented by a Venn diagram.

18.3 Ontological Pyramids

The relation $<_o$ of ontological precedence may be characterized thus: $A <_o B$ if B is about objects constituted by individuals studied by A.

For example, physics precedes chemistry ontologically because it concerns molecules, whose elementary constituents are studied by physicists. And biology is ontologically based on chemistry, because it studies cells, which are composed of molecules. But chemistry does not precede biology in the strong sense, because to perform a chemical analysis of a cell one must start by characterizing the cell in biological terms—e.g., by its metabolism and reproduction.

The psychology-sociology pair has been the object of vehement but inconclusive debates: whereas holists hold that the whole precedes its parts, whence sociology precedes psychology, individualists maintain the opposite. I suggest that this debate can be resolved by using the above distinctions among different kinds o precedence. Psychology precedes sociology in the ontological sense because the social groups, which are the referents of social science, are constituted by individuals endowed with mental faculties. But from this it does not follow that psychology entails sociology either epistemolgically or logically, because the social structure, which is a global or systemic property, must be known to account for individual behavior.

The relation $<_o$ of ontological precedente induces induce pyramids like this one:

SOCIAL SCIENCE
PSYCHOLOGY
BIOLOGY
CHEMISTRY
PHYSICS

Note once again that the pyramidal arrangement ignores the hybrid sciences, which can easily be accommodated in a rosette. Note also that, although all the scientific disciplines include all the mathematics they can chew, mathematics does not occur in the above pyramid. The reason is of course that the mathematical objects are imaginary or fictious, not material.

18.4 Place of the Universal Sciences

What is the place of the two universal disciplines, mathematics and philosophy, in the knowledge system? It is usually held that philosophy and mathematics are either at the base or at the top of the knowledge pyramid. Let us start with philosophy.

I have argued elsewhere (e.g., Bunge 1998) that all of the factual ("empirical") sciences certain ontological principles, such as that reality is consists exclusively of concrete entities, and that none of these is isolated from the rest.

If this thesis of the partial intersection of science and philosophy is accepted, we must also admit that some philosophy precedes epistemologically all the fields of factual scientific research. But, of course, philosophy holds no such pride of place in the remaining pyramids. In particular, philosophy does not occur in the pyramid induced by the relation of logical precedence, because those principles do not imply anything and because, for all their important, they make up only a small part of philosophy.

Something similar is the case with the other universal discipline, namely mathematics (which has logic at the base). Every factual science embodies, at any given time, only a small part of mathematics; but it may use any branch of it, depending on the knowledge and ingenuity of the investigator. Hence mathematics, or rather some of it, may be placed at the base of any epistemological pyramid.

However, the pyramid metaphor is not altogether suitable to place either mathematics or philosophy, because it fails to suggest that parts of those disciplines are included, at least potentially, in all the sciences (and technologies) of reality. This idea, by contrast, is suggested by the rosetta metaphor. In fact, the latter is a graphic representation of the concept of intersection or logical product of disciplines (or rather the corresponding sets of propositions).

More precisely, the philosophical and mathematical nuclei of the factual sciences and technologies may be characterized as what those disciplines have in common, that is, their intersection. In other words, the rosette of the sciences has two foci: a fragment of mathematics and another of philosophy. In turn, these foci overlap partially: their intersection is exact philosophy (Bunge 1972).

The rosette arrangement also helps understand the various legal philosophies. According to the Natural Law doctrines (in both its secular and religious versions), natural law is included in morals. By contrast, legal positivism claims the purity (moral and political neutrality) of law, as well as its autonomy with respect to science. (The law would be the only technology without a scientific base.) Finally, legal realism places the law in the intersection of politics, morals, and social science. Thus, according to it, crime is definable as the social action that breaches positive laws that entrench moral rules.

18.5 Concluding Remarks

The preceding analysis may be summed up as follows.

1. The bodies of knowledge may be arranged into pyramids of different types. The ambiguous expressions of the form "B is based on A" may easily be disambiguated upon analyzing "is based on" into three different precedence relations: logical, epistemological, and ontological.
2. None of the precedence relations defined above coincides with that of historical precedente. For example, mathematical logic and set theory, usually regarded as the foundations of mathematics, emerged nearly three millennia later—a case of building starting with the roof.

3. The pyramidal arrangements make no room for the hybrid disciplines, such as socioeconomics and the law. These are easily placed in rosettes or Venn diagrams. Moreover, this arrangement suggests the unity of knowledge, which is shown as a system rather than a collection of disjoin disciplines. Note, however, that religion and pseudoscience remain outside any knowledge rosette—which suggests that they are not bodies of knowledge.
4. In arranging the bodies of knowledge on the same footing, the rosette metaphor misses the important ideas of logical, epistemological, and ontological precedence. Consequently, it is advisable to use concurrently both metaphors: one to emphasize precedente, and the other to emphasize the partial overlap and resulting unity of the bodies of knowledge.

References

Bunge, Mario (ed.). 1972. *Exact philosophy*. Dordrecht: Reídel.
Bunge, Mario. 1974. *Treatise on basic philosophy*, vol. 1 and 2. Dordrecht: Reídel.
Bunge, Mario, ed. 1998 [1967]. *Philosophy of science*, 2 vols. New Brunswick: Transaction.
Bunge, Mario. 2003. *Emergence and convergence*. Toronto: University of Toronto Press.

Chapter 19
Existence: Single or Double?

Presumably most people distinguish between real (or de re) existence and ideal (or de dicto) existence. For instance, we normally distinguish roads from road maps. This distinction is so fundamental, that those who fail to draw it are diagnosed as schizophrenics in daily life, and as logicians in academia. In fact, a logical imperialist, such as Tarski or Quine, claims that there is a single existence concept, valid in physics as well as in mathematics, namely, the existential quantifier. Who is right in this case: commonsense or the sophisticated mathematical logicians? Let us see.

19.1 Existence and Someness

The so-called "existential" quantifier $\exists$ occurs as a prefix in many mathematical formulas, such as the equation $\exists x(x^2 - 1 = 0)$, where x designates a number. In their *Principia Mathematica* (10.01), Whitehead and Russell defined $\exists$ in terms of negation and the universal quantifier, namely as not-all-not. That is, when asserting $\exists xPx$ one only says that not all individuals lack the property P.

However, the formula $\exists xPx$ is usually read as the statement that *there are* individuals endowed with P. Yet, this interpretation does not fit the definition of the quantifier and is at variance with usage. For example, an atheist will deny that there are angels, but he should not object to the assertion that some angels are guardian angels. Existence entails someness, not the other way round.

Furthermore, Quine (1953: 12) has emphatically and repeatedly claimed that the interpretation of $\exists$ as existence proves that logic, far from being topic neutral, has an ontological commitment. I hold that, on the contrary, this thesis overflows logic: that the above-cited definition shows that the sober reading of $\exists$ es "some". In other words the correct reading of "$\exists xPx$" is not "There exist Ps", but "Some individual are Ps." If I am right, we have to talk of *someness*, not of *existence*. True, 'some' is imprecise, but 'exist' is ambiguous as long as no mention is made of the universe where the individual(s) in question lodge. In other words, existence simpliciter is contextual.

M. Bunge, *Evaluating Philosophies*, Boston Studies in the Philosophy of Science 295, DOI 10.1007/978-94-007-4408-0_19, © Springer Science+Business Media Dordrecht 2012

The prefix $\exists x$ to the equation $x^2 - 1 = 0$ encourages us to try and solve it. If we do so, we find that it has two roots or solutions, +1 and −1. In traditional language, one says that the individual *x is* [equal to] 1, or *is* [equal to] −1. This idiom induces the illusion that, when solving that equation, one passes from the indeterminate to the determinate, from potency to act, from possibility to existence. Maybe this is a root of the famous semantic aphorism *To be is to be the value of a bound variable.*

This aphorism, that may be taken as the emblem of logical imperialism, overlooks the radical difference between real or material existence, on the one hand, and ideal or formal existence, on the other. This difference explains why, whereas the proofs of formal existence are purely conceptual, those of real existence are empirical. For example, several thousands physicists are currently making delicate measurements to try and find evidence for the real existence of gravitons, Higgs bosons, and other entities hypothesized by the theorists. Do not tell them that there is a single existence concept, for hey know that there is a chasm between existence on paper and real existence.

If the said difference is admitted, and if logic is regarded as the common basis of all the sciences, both formal and factual, it follows that logic must be both baseless and ontologically neutral rather than committed. A materialist may admit that some "worlds," such as those of mathematics and science fiction, are imaginary and therefore immaterial, without assenting to Plato's thesis that they exist by themselves, outside imagination.

19.2 Existential Predicate and Anselm's Proof

Although logicians have not tired of repeating that existence is not a predicate, it is possible to construct an existence predicate. Here it is. Let C design a well-defined subset of a set X. That is, C is the extension of a precise predicate, such as "is even", "is opaque", and "is father of." Call χ_C the characteristic function of C. This is the function $\chi_C: X \rightarrow \{0,1\}$ such that $\chi_C(x) = 1$ if x is in C, and 0 otherwise. Let us agree that an individual x *exists* in C if and only if $\chi_C(x) = 1$. The (contextual) existence predicate E_C is then the function E_C: $C \rightarrow$ Set of the propositions including E_C. For example, the proposition $E_C b$ will be read "b exists in C." If C = Reality, the existence in question is real; otherwise it is imaginary. (To accommodate the ideal existence of sets, as in the axiom of choice, C must be assumed to be a category, not a set.)

Thus, to be is to belong to one of the two basic categories: that of material things, and that of mind-made ones. The material/ideal (or concrete/abstract, or real/imaginary) dichotomy was examined in Chap. 14. Suffice it here to say that, whereas the material objects are changeable, the ideal ones do not change by themselves since they are mind-made, although they are not mental. (In a materialist perspective, mental objects are brain events, whereas abstract objects are fictions.)

Let us now use the existential predicate introduced above to revisit the most famous of all the arguments for God's existence. Anselm of Canterbury argued that

God exists because He is perfect, and existence is a property of perfection. Some mathematical logicians have claimed that Anselm was wrong because existence is not a predicate but the $\exists$ quantifier. I suggest that this objection is sophistic because in all the fields of knowledge we tacitly use an existential predicate that has nothing to do with the "existential" quantifier, as when it is asserted or denied that there are living beings in Mars or perpetual motion machines.

Using the existence predicate defined a while ago, we may reformulate Anselm's argument as follows.

God is perfect	Pg
Everthying perfect exists in R [really]	$\forall x(Px \Rightarrow E_R x)$
God exists in R.	$E_R g$

Both premises are controversial, particularly the first one, since it presupposes the existence of God. Hence the atheist will have to propose serious arguments against it instead of the sophistry of the logical imperialist. An alternative is to admit the existence of God for the sake of argument, and add the ontological postulate that everything real is imperfect: that if something is perfect then it is ideal, like Pythagoras' theorem or a Beethoven sonata. But the conjunction of both postulates implies the unreality of God. In short, Anselm was far less wrong than his modern critics would have it.

19.3 Logic and Ontology

The conflation between formal and real existence suggests that logic deals with the world, and even that it may study facts without looking at them. An instance of this plan of invasion and conquest of the factual sciences by logic is the attempt to eliminate the extralogical predicates or, equivalently, to replace the natural laws with definitions (Quine and Goodman 1940). Do not panic: the alleged proof is fallacious, for it involves a key formula taken out of the blue, as if symbols could be assigned meanings out of context (Bunge 1998: I, 148–149).

Maybe Quine's celebrated thesis about the ontological commitment of logic stems from another confusion, namely the equation of 'reference class' (or 'universe of discourse') with 'ontology'—parallel to the popular confusion between environment and ecology. Let us clarify this point with an example. Religiosity can be attributed only to human beings. That is, the reference class of the predicate "is religious" is humankind. Quine would say that this is the *ontology* of the predicate in question. But this usage of 'ontology' is incorrect, because every ontology is a theory, that is, a hypothetic-deductive system, not the domain of a predicate. Actually, the ontological commitment clause may be taken as nothing but the condition that the universe of discourse be non-empty. And this calls existentialism to mind.

The distinction between existence (E_C) and someness (∃) applies to the objection to existentialism for treating existence as a predicate (Edwards 2004). The existentialists and Hamlet are right in holding that "to be or not to be: that is the question." But of course they are wrong in treating the problems of human existence as a mere matter of words, such as *Dasein*, that are hermetic and therefore outside logic and science—just as Heidegger intended.

The confusions between reality and fiction, and between existence and someness, are inherent in what we have called *panlogicism* or *logical imperialism.* This is the belief that logic is both necessary and sufficient to philosophize. Whoever adopts this point of view believes to be in a position of approaching all the epistemic problems with the sole help of logic.

This strategy impoverishes almost as much as irrationalism, because it discards all the sciences and technologies. Logic is a very potent tool precisely because, since it does not commit itself to any view of reality, it can be used to analyze any ideas whatever, regardless of their referents. In other words, the logical truths refer to anything, hence they describe nothing extra-logical in particular (Bunge 1974a, b). If we abuse of logic, entrusting it with tasks it cannot perform, we will discredit it, as happened with the late schoolmen, who pontificated logically about nature without making any observations, much less experiments. It took logic three centuries to recover the prestige it lost during the Scientific Revolution, when Francis Bacon and others accused it of being barren.

19.4 Truth: Single or Double?

If there a single kind of object and existence, there must also be a single kind of truth. This thesis may be dubbed *alethic monism*. The objective idealists, who hold that all existents are ideal, will claim that truth is an idea-idea correspondence This is the *coherence* theory of truth. By contrast, the nominalists, who abhor the idea of an idea, hold that truth is a sign-sign or sign-fact correspondence, and will accordingly claim that the *correspondence* theory of truth fits both mathematics and factual science. This was the view of Buridan, Hobbes, Hilbert, and Tarski—no less. Both versions of alethic monism are consistent, but neither of them is clear and adequate. I have criticized them I some detail elsewhere (Bunge 1974b). Here I will confine myself to the main points.

Consider the following statement

$$\text{7 is a prime number.} \tag{19.1}$$

What does this statement refer to? Obviously, it is about numbers and nothing else. And is it true? Yes, for 7 satisfies the definition of a prime number. So, the truth of (19.1) is formal, not factual: it is a matter of coherence (among constructs), not of correspondence of construct to fact.

Let us now examine the formula

$$\exists x\left(x^2 + x = 0\right) \tag{19.2}$$

This condition on the otherwise unspecified individual x (the unknown) is satisfied by the number −1, as well as by the negative of a unit matrix I of any order: 2×2, 3×3, etc. That is, (19.2) is true for the valuations or assignments x:=−1 and x:=−I. But by itself, (19.2) refers to nothing in particular, and as a consequence it is neither true nor false. When the variable *x* is interpreted as a certain number, or as a matrix of a certain kind, does (19.2) become a proposition proper, and as such one with a (formal) truth value. This is the notion of truth as *satisfaction* in a certain domain of mathematical objects (numbers, matrices, or what have you).

This is the concept that Alfred Tarski (1956) defined in his paper on truth in "formalized languages"—his nominalist name for abstract theories, such as set theory and group theory. Truth as satisfaction is clearly a case of truth by coherence. This concept of truth suffices for abstract theories but not for interpreted mathematical theories, such as Euclidean geometry, number theory, and the infinitesimal calculus: in these theories, there is also truth by definition and by proof.

In his celebrated philosophical paper Tarski (1944), offered a second concept of truth, which he regarded as the precise formulation of the well-known if imprecise notion of truth as correspondence (*adaequatio intellectus ad rem*). It boils down to the following schema:

s' is a true sentence if and only if s. T

This somewhat cryptic thesis has been the object of indecisive and unproductive debate for seven decades. What is clear is that, since *T* only relates sentences to their names—a bit of language to its metalanguage—it remains on the linguistic (or conceptual) level, whereas genuine "correspondence" relates the symbolic to reality. Thus, Tarski was mistaken in believing that he had defined the classical concept of factual truth as the adequacy of idea to fact.

When people seek data or put factual hypotheses to empirical tests, they focus on <proposition, fact> pairs, not <sentence-sentence name> pairs. Let us examine a couple of examples, starting with the statement

There are smart children. (19.3)

This statement refers to children, hence to check for its truth we have to test children, not numbers or equations. The result is well known: (19.3) fits the facts, it is factually true.

Now consider the centerpiece of the classical mechanics of point particles:

For all point particles, relative to all inertial frames, and at all times, the force on a particle equals its mass times its acceleration:

$$F = md^2x/dt^2. \tag{19.4}$$

This formula refers to material objects of a certain type, and it contains five extralogical predicates: reference frame, force, mass, position, and acceleration. Every one of them represents a physical item. So, although (19.4) is a mathematical formula, it is usually interpreted in such a manner that it represents a piece of reality, namely a law of motion of material objects of a kind.

The said interpretation, along with certain indicator formulas, allows one to check for the factual truth of (19.4) through empirical tests. We know the verdict: (19.4) holds only for medium-sized bodies moving at small speeds compared with the velocity of light in a vacuum. Thus (19.4) is *approximately* true to fact. In particular cases it is possible to calculate the error, or departure from truth, of formulas in a given scientific theory relative to their counterparts in different theories. This comparison has been done for Galileo's and Einstein' laws of falling bodies (Bunge 1998: II, 344).

The investigators in all the sciences and technologies know that they work, at best, with approximate truths. This is why they have no use for the concept of truth proposed by Tarski, it is utterly irrelevant to science and technology, because (a) it ignores the semantic concept of reference; (b) it conflates two kinds of truth; and (c) it ignores the concept of partial truth, that is central to approximation theory as well as to the methodology of the factual sciences. Why is it still being widely regarded as the solution to the problem of truth? And what does this fact tell us about the current state of philosophy?

19.5 Conclusion

There are two kinds of existence: real (or material) and ideal (or imaginary). This important distinction is lost when the "existential" quantifier ∃ is interpreted as "exists" instead of "something". And this confusion induces another two, namely the theses that logic presupposes ontology, and that there is a single kind of truth. Logic is potent but not omnipotent, Moreover, logic is a stellar servant but a bad boss.

References

Bunge, Mario. 1974a. The relations of logic and semantics to ontology. *Journal of Philosophical Logic* 3: 195–210.

Bunge, Mario. 1974b. *Treatise on basic philosophy, vol. 2: Interpretation and truth*. Dordrecht: Reidel.

Bunge, Mario. 1977. *Treatise on basic philosophy, vol. 3: The furniture of the world*. Dordrecht: Reidel.

Bunge, Mario. 1998 [1967]. *Philosophy of science, 2 vols*. New Brunswick: Transaction.

Edwards, Paul. 2004. *Heidegger's confusions*. Amherst: Prometheus Books.

Quine, Willard Van Orman. 1953. *From a logical point of view.* Cambridge, MA: Harvard University Press.

Quine, Willard Van Orman, and Goodman Nelson. 1940. Elimination of extra-logical predicates. *Journal of Symbolic Logic* 5: 104–109.

Tarski, Alfred. 1956. The concept of truth in formalized languages [1933]. In *Logic, semantics, metamathematics*, 152–278. Oxford: Clarendon Press.

Tarski, Alfred. 1944. The semantic conception of truth and the foundations of semantics. *Philosophy and Phenomenlogical Research* iv: 341–376.

Chapter 20
Conclusion: Valuation Criterion

Every research project, in any field, consists in posing some problem, old or new, and in attempting to solve it. These tasks are guided by antecedent knowledge, intuition, and value judgment: one assesses the worth of the problem, the means to handle it, and the possible solutions.

Those who do not tackle new problems, and are therefore content with using some routine, can afford to ignore all value-related scruples. But which valuation criterion should pick those who attack philosophical problems requiring original research—original ideas, not just literature search—and therefore promise new findings? Let us tackle this problem.

20.1 Monism or Pluralism?

It is well known that, although they may differ in their referents, all the sciences are methodologically one: all of them seek clarity, rigor, generality, coherence, depth and, above all, truths, even if approximate. (See Mirowski 2011 for a criticism of the neoliberal thesis that the aim of science is to make money.)

What a contrast between the harmony of the sciences and the cacophony of philosophies! Every philosopher defines his field at his pleasure, so he may disqualify out of hand the doctrines he dislikes. Only the philosophy professors lacking in convictions will treat all the schools with equanimity: they neither love nor hate any of them. Theirs is a routine job, such as that of tending graves, not the realization of an avocation.

What is the researcher to do when faced with the multiplicity of philosophies? The answer depends on whether the aim is to enrich philosophy or to get along with colleagues of views radically different from one's own. As Nicholas Rescher (1985) says, the individual is inherently monistic, whereas the community is inherently pluralistic—at least in a democratic society. We need critics to stay awake, just as we should fight censorship because it stifles creativity.

M. Bunge, *Evaluating Philosophies*, Boston Studies in the Philosophy of Science 295, DOI 10.1007/978-94-007-4408-0_20, © Springer Science+Business Media Dordrecht 2012

The monism of the individual researcher is due to the fact that he struggles with problems, and seeks to find *one* solution, the best, to every one of them. In other words, whoever works on a research problem may start by doubting, but he ends up by taking a position, that is, adopting a solution to his problem—which may well be that it is insoluble. Only the passive onlooker can afford to dispense with valuation criteria.

20.2 The Fertility Criterion

The preceding pages suggest an objective yardstick to measure the worth of philosophies: *By their fruits ye shalt know them: Tell me what your philosophy is doing for the search for truth or the good, and I will tell me what it is worth.*

According to this evaluation criterion, the irrationalist and subjectivist schools are barren in the best of cases, and harmful in the worst, whereas the others are either beneficial to some extent or in some context, or indifferent to the exploration of the real world or of the domain of ideas. Let us peek at two examples separated by six centuries: Thomism and phenomenology. The Catholic Church embraced the former (after having banned it for its heterodoxies) until about the middle of the last century, when it briefly considered adopting phenomenology for being idealist and conservative, as well as for looking modern.

The philosophy of Thomas Aquinas was progressive in its time because it exalted realism and rational debate, as well as the study of nature, whereas its irrationalist rival, Augustine of Hippo's, had discarded such study alleging that it helped neither self-knowledge, which is false, nor salvation, which is true.

By contrast, Edmund Husserl's phenomenology is utterly regressive in being cryptic, in urging detachment from reality, and in rejecting science while claiming to be "rigorous science." Despite being ego-centered, phenomenology has made no discoveries about the self. For example, Husserl defined self-consciousness as "the absolute being," which invites defining the unconscious as "the relative being", an equally hermetic expression and therefore utterly useless in psychology and in medicine. Worse, the German original, "Bewusst*sein* ist absolutes *Sein*", is a wordplay. What does it mean? And how does it help the neurologist or the anesthesiologist who needs to know whether that patient, who looks comatose, perceives external simuli and thinks about herself?

And what can be said about Martin Heidegger, Husserl's star pupil and successor, and his many imitators, such as Jean-Paul Sartre and Jacques Derrida? That they did not solve any philosophical problems, because they only wrote sentences that were either trivial or nonsensical. For example, Heidegger became famous just because he wrote idiocies such as *die Welt weltet* (the world worlds); Jean-Paul Sartre, like Heidegger, wrote about nothingness as if "this" could have properties; and Jacques Derrida gained celebrity writing *rien hors du texte*, a French version of Heidegger's (1976: 11) sentence "Im Wort, in der Sprache werden und sind erst die Dinge" ("Things are and become only in the word, in language.")

What have those scribblers contributed to knowledge? Just as much a Nietzsche or the Zen Buddhists intent on "emptying the mind": nothing. They succeeded in *épater le bourgeois* without offering serious alternatives to the so called official thinking. Worse, those authors and their successors, the self-styled postmoderns, have been denouncing the most advanced, intelligent, disinterested, and yet useful of all the accomplishments of modernity, namely science.

20.3 Conclusion

Our evaluation criterion is not a neutral arbiter, since it presupposes that progress is at once possible and desirable, and that morals and praxis count. The adoption of that criterion does not involve restricting philosophical pluralism, although it does involve discarding meaningless verbiage passed off as philosophy. Pluralism is necessary to dialogue with oneself and with colleagues, to learn, to clarify ideas, to question, to resolve doubts, and to solve problems.

But discussion is only a means: the aim of every research project is to pose and solve some problem, even knowing that new findings may correct the proposed solution or even discard the corresponding problem. Whence a second metaphilosophical norm: Seek the unity of your own philosophy but tolerate the diversity of all authentic philosophies, and promote the rational debate among them. Few things are more discouraging than being right without anyone noticing.

References

Heidegger, Martin. 1976 [1953]. *Einführung in die Metaphysik*, 5th ed. Tübingen: Max Niemeyer.

Mirowski, Philip. 2011. *Science-mart: Privatizing American science*. Cambridge, MA: Harvard University Press.

Rescher, Nicholas. 1985. *The strife of systems*. Pittsburgh, PA: University of Pittsburgh Press.

Philosophical Glossary

Axiology Value theory

Background knowledge What is known so far, the starting point of a research project.

Computationism The programmatic hypothesis according to which natural or social processes of certain kinds (e.g., mental) are computational or algorithmic. Radical computationism holds that to be is to compute or to be computed.

Construct Conceptual object: concept, proposition, norm, classification, or theory.

Constructivism Ontological: The world is a human (individual or social) construction. Epistemological: all ideas are constructs, and none derive directly from perception.

Counter-Enlightenment Reaction against the French Enlightenment. Targets: rationality, scientism, materialism, and democracy. Icons: Hegel, Nietzsche, and Heidegger.

Counter-example Exception to an alleged law or rule.

Datum A piece of information, such as "She phoned". A scientific datum is one obtained with the help of a scientific technique, and that can either confirm or infirm a hypothesis.

Deontologism The moral philosophy that binds us to our duties regardless of consequences.

Dialectics The ontological doctrine, due to Hegel and adopted by Marx and his followers, according to which every item is at once the unity and struggle of opposites.

Dualism The family of doctrines according to which there are two equally basic but mutually irreducible kinds, such as matter and mind, or nature and society. A particular case of pluralism.

Emergence The advent of qualitative novelty. A property of systems.

Empiricism The epistemological doctrine according to which experience is the only source and test of all ideas.

Energy Changeability, the universal property of concrete things.

Enlightenment The eighteenth century intellectual movement that exalted reason, the search for truth, and human rights.

M. Bunge, *Evaluating Philosophies*, Boston Studies in the Philosophy of Science 295, DOI 10.1007/978-94-007-4408-0, © Springer Science+Business Media Dordrecht 2012

Epistemology The philosophical study of cognition and its product, knowledge. Main schools: skepticism, intuitionism, empiricism, conventionalism, fictionism, rationalism, and ratio-empiricism.

Event Change of state of a concrete thing.

Ethics The philosophical study of morality, in particular moral norms.

Evidence A set of data relevant to some hypothesis.

Existence The main property of an object. Conceptual: belonging to a conceptual system, such as a theory. Material: being a part of the real world, or having energy.

Existential quantifier Formalization of the concept of someness, as in $\exists xPx$, or some individuals have the property P. Often wrongly equated with the concept of existence.

Explanation Description of a mechanism.

Fact State or change of state of a material thing.

Functionalism The view that only function matters, whereas "substrate" (matter) does not.

Hermeneutics The school of thought according to which (a) symbols are all-important or even the only existents, and (b) all social facts are to be understood through the intuitive capture of the actors' intentions.

Holism The family of doctrines according to which all things come in unanalyzable wholes. The ontological partner of intuitionism. Antonym: individualism.

Hypothesis A proposition that asserts more than any data relevant to it.

Idealism, philosophical The family of ontologies according to which ideas pre-exist and dominate everything else. Two main varieties: subjectivism (e.g., Berkeley and Kant) and objective realism (e.g., Plato and Hegel.

Individualism The view that the universe is an aggregate of separate individuals: that wholes and emergence are illusory.

Induction The leap from a set of particular propositions to a universal proposition.

Inductivism The thesis that all valid universal propositions are inductions.

Interdiscipline A discipline formed by the fusion of two or more disciplines. Examples: cognitive neuroscience and socioeconomics.

Intuition Spontaneous (preanalytic) insight.

Intuitionism The family of epistemological doctrines that assert the primacy of intuition (preanalytic insight). The epistemological partner of holism.

Law Regularity or pattern. Objective law: inherent in things, hence a property of theirs. Law statement: formula that captures an objective pattern. Sometimes called 'nomological statement' to avoid confusion with its referent.

Lawfulness, principle of The philosophical hypothesis, presupposed by all the sciences, according to which everything, even chance, satisfies laws.

Level (of organization) Collection of things characterized by a bundle of properties. Examples: physical, chemical, biological, social, semiotic, and technological levels.

Logic The theory of deduction, inherent in all the disciplines.

Macro-reduction Reduction to higher-level entities or processes. Example: explanation of individual behavior in terms of her social standing.

Material Capable of changing by itself, having energy.

Materialism, philosophical The family of naturalist ontologies according to which all existents are material. Three main versions: vulgar (or physicalist), dialectical, and emergentist (or systemic). Physicalism asserts that everything is physical. Dialectical (or Marxist) materialism is the fusion of materialism with dialectics: Every existent is at once material and the unity of opposites. And emergentist (or systemic) materialism is the fusion of materialism with systemism: Every existent is at once material and a system or part of one.

Matter The set of all material things.

Meaning, semantic a property of constructs. Reference or denotation together with sense or connotation. Not to be confused with the goal of an action.

Mechanism The totality of processes that make a system tick, such a metabolism in the case of organisms, and trade in that of markets.

Metaphysics *See* Ontology

Method, scientific Background knowledge → Problem → Hypothesis or technique (special method) → Test → Evaluation → Revision of background knowledge → New problem.

Methodology The part of epistemology that studies methods.

Micro-reduction Reduction to lower-level entities. Example: rational-choice theories.

Mind Primitive view: The immaterial soul, or else the set of mental faculties or functions, that may be understood without the help of neuroscience. Scientific view: The family of specific functions of highly developed brains, whence it can only be explained by cognitive and affective neuroscience.

Mind-body problem The question of the nature of the mental functions and their relation to the brain.

Naturalism The family of ontologies that assert that all existents are natural–hence none are supernatural.

Nominalism Medieval doctrine that denies the existence of ideas and replaces them with linguistic items (e.g., 'sentence' instead or 'proposition').

Objectivism The view that it is possible and desirable to describe all items in the external world without reference to the subject. Syn.: realism.

Ontology The philosophical study of being and becoming

Phenomenalism The philosophical view that there are only phenomena (appearances to someone), or that only these can be known. Phenomenalism is anthropocentric and it opposes realism.

Phenomenon Perception of a fact, in contradistinction to the fact in Itself. For instance, color as distinct from wavelength. Phenomena are real but they happen only in brains.

Pluralism The family of philosophies that assert the plurality of basic kinds (substance pluralism), basic properties (property pluralism), or both. Emergentist (or systemic) materialism is property-pluralist but substance-monist: one substance (matter), many properties.

Positivism Modern empiricism. Classical positivism (Comte, Mill, Mach) was phenomenalist, inductivist and diffident of theory. Logical positivism (Schlick, Carnap, Reichenbach) was phenomenalist, attempted to reduce the theoretical to the empirical, and stressed the importance of modern logic.

Postmodernism The rejection of the intellectual values of the French Enlightenment, in particular clarity, consistency, and objective truth. A combination of hermeneutics with irrationalism.

Pragmatism The philosophical doctrine according to which practice is the source, measure, and goal of all knowledge and all value.

Praxiology Action theory.

Predicate Conceptualization of a property. Predicates can be unary, like "is young", binary, like "older than", ternary, like "between", and in general *n*-ary.

Process A non-instantaneous change of state of a thing: a sequence of states of a thing.

Property Feature or trait of an item, such as the length of a segment, the energy of a bullet, the age of an organism, an the structure of a social group. There are neither properties without bearers, nor objects without properties.

Proposition A statement, such as "You are reading", capable of being true or false to some degree. Any given statement can be expressed by a number of different sentences.

Pseudoscience A body of belief or a practice sold as scientific although it is either untestable or at variance with the bulk of extant knowledge.

Ratio-empiricism Merger of rationalism and empiricism. Examples: Kant, logical empiricism, and scientific realism.

Rationalism The epistemological doctrine according to which cognition involves reasoning. Radical or aprioristic rationalism: reason is sufficient for knowing. Moderate rationalism: reason is necessary for understanding.

Real Existent in and by itself, that is, whether or not it is perceived or conceived by someone. Idealism holds that ideas are real; materialism that only material items are real.

Realism The philosophy according to which the external world exists independently of the inquirer, and can be known at least in part. Scientific realism is the variety of idealism that takes scientific research to be the best means to acquire knowledge.

Reduction Conceptual operation whereby an item is shown to be identical with another or included in it, or to be either an aggregate, a combination, or an average of other items.

Reductionism Research strategy that attempts to understand complex things to or their composition.

Rule Standardized prescription for doing something, or human behavior pattern.

Science The branch of knowledge characterized by rationality, testability, and systemicity. Science is split into two great domains: formal (logic and mathematics), and factual–natural, biosocial, an social.

Scientism The epistemological thesis that whatever can be known is best investigated scientifically. Not to be confused with the disregard for non-physical features.

Semantics The study of meaning and truth. Not to be confused with 'a mere matter of words'.

Skepticism The family of epistemologies that either deny the possibility of certain knowledge (radical skepticism) or else affirm that we should doubt and check before believing–and often even after having accepted a belief (moderate skepticism).

State (of a thing) The totality of properties of a thing at a given instant. Example: the UN human development index.

State space The set of al the possible states of a thing at a given instant.

Structure of a system The totality of relations among the system's components, and among these and the system's environment.

Subjective Something happening inside the subject's head, such as feeling elated or sad.

Subjectivism The family of philosophies according to which everything is in a subject's mind (subjective idealism), or nothing can be known objectively. Antonym: 'objectivism'.

System A complex object whose constituents are held together by strong bonds—logical, physical, biological, or social—and possessing global (emergent) properties that their parts lack.

Systemics The branch of ontology concerned with the features common to all concrete systems

Systemism Ontology: Everything is either a system or a component of some system. Epistemology: Every piece of knowledge is or ought to become a member of a conceptual system, such as a theory. Axiology: Every value is or ought to become a component of a system of interrelated values.

Tautology A truth of logic, such as "People rebound because they are resilient".

Technology Body of practical knowledge invented and utilized to design, produce, or maintain artifacts, whether physical (like computers), biological (like farm animals), or social (like schools),

Testability, empirical The ability of a hypothesis or a theory to be either confirmed or falsified to some extent by empirical data. A characteristic of science and scientific philosophy in contradistinction to theology and esotericism.

Theory Hypothetico-deductive system: A system of propositions every one of which implies or is implied by others.

Truth, factual The adequacy of an idea to the facts it represents, as with "It rains" when it is actually raining.

Truth, formal The coherence of an idea with a previously accepted body of ideas. Examples: tautologies and theorems.

Truth, partial The less than perfect adequacy of an idea to its referent, as in "$\pi = 3$."

Truth value One of he various degrees of adequacy (true, false, half- true, etc.) that a proposition may have or acquire.

Universal quantifier Prefix or operator "for all", as in "$\forall x$(x is real if and only if x is material)."

Utilitarianism The moral philosophy that enjoins us to maximize the expected utility or gain for self or others. Rivals: deontologism ("Do your duty regardless of consequences") and agathonism ("Seek the good for self or others").

Utility Subjective value.

Verstehen (interpretation) Intuitive understanding. The "method" of hermeneutics.

Index

A

Abduction. *See* Guessing
Abstraction, 163
Accident, 22, 88, 106, 156. *See also* Chance
Action, 18, 25, 27, 30, 36–38, 46, 52, 87, 90, 106, 125, 126, 145, 159, 160, 170, 187, 188
Agassi, J.,
Agonism, 86, 88
AIDS, 156
Albert, H., 29
Alethic monism, 176
Algorithm, 42, 62, 116
Altruism,
Analogy, 135, 159
Analysis, 8, 10, 11, 21, 24, 37, 38, 86, 123, 127, 133, 143, 169, 170
Anselm's proof, 174–175
Antipsychiatry, 164
Antirealism,
Appearance, 18, 31. *See also* Phenomenon
Approach, 4, 5, 9, 18, 21, 24, 30, 36–46, 163, 164
Approximate truth, 178, 181
Approximation, 15, 106, 178
Apriorism, 4, 25, 89–91, 115
Aquinas, T., 19, 182
Archaeology, 24, 30, 41, 52, 116
Aristotelianism, 3, 12, 13, 19
Aristotle, 7, 12, 17, 19, 20, 25, 44, 52, 55, 85, 86, 89, 135, 163
Article 48, 100
Artifact, 24, 38, 43, 54, 59, 63, 64, 87, 111, 116, 154, 155
As if, 8, 45, 63, 77, 121, 164, 175, 182. *See also* Fictionism
Aspect, A., 146
Astronomy, 18, 116
Atomism, 29
Augustine, 13, 66, 182
Aumann, R.J., 78, 79
Axiology, 185, 189
Axiom, 109, 120, 174
Axiomatization,

B

Background knowledge, 15, 37, 185, 187
Bacon, F., 19, 33, 51, 88, 115, 176
Barraclough, G., 30, 90
Barrow, J.D., 154
Bayesian, 105–107, 109, 110, 112, 116–118
Bayesian induction, 116–118
Bayesianism, 105, 106, 108, 110–111
Bayes' theorem, 110, 117
Becker, G.S., 28, 72, 87
Belief, 10, 15, 22, 32, 41, 67, 69, 88, 104, 107, 109–111, 118, 137, 139, 147, 152, 159, 168, 176, 188
Bell's theorem, 145
Benedict XVI, 16, 83
Bergson, H., 6, 8, 13, 29
Berkeley, G., 6–8, 112, 139, 144
Beynon-Davies, P., 61
Biology, 6, 12, 17, 21, 22, 31, 40, 41, 44, 55, 61, 62, 86, 89, 106, 123, 124, 133, 160, 167–169
Biosocial science, 8, 42–43, 106, 167
Biostatistics,
Bit, 5
Black bo,
Blanc, L., 69
Bohm, D., 145

M. Bunge, *Evaluating Philosophies*, Boston Studies in the Philosophy of Science 295, DOI 10.1007/978-94-007-4408-0, © Springer Science+Business Media Dordrecht 2012

Bohr, N., 6, 19, 29, 127, 131, 139, 145, 148
Bok, D., 66
Born, M., 139
Bouchaud, J.-P., 81
Braudel, F., 41
Bressler, S., 161
Bringezu, S., 59
Bruera, J., 101
Buddhism, 188
Buonomano, D.V., 162
Buss, D.M., 152

C
Capitalism, 13, 18, 24, 33, 69, 70, 78, 83, 87, 90–92
Carnap, R., 20, 27, 116, 119, 120, 160, 187
Catholic Church, 182
Causalism, 108
Causality,
Causation, 23, 27, 133, 147
Censorship, 17, 181
CESM model of a system, 39
Chaitin, G., 154
Challenger space shuttle, 110, 111
Chance, 23, 67, 100, 106–109, 133, 147, 186. *See also* Accident; Disorder; Probability; Randomness
Changeability, 21, 133, 137, 185. *See also* Energy; Material
Chemistry, 6, 21, 44, 77, 106, 115, 124, 126, 132, 140, 167–169
China, 17, 58, 73, 86
Chomsky, N., 16, 163
Cini, M., 143
Clarity, conceptual, 5, 13, 23, 24, 26, 27, 39, 42, 61, 86, 107, 120, 121, 134, 146, 174, 177, 181, 185, 186, 188, 189
Class, 17, 69, 71, 72, 84, 86–89, 91, 144, 175. *See also* Kind; Species
Classical physics, 40, 124, 127, 134, 142, 143, 148
Classicism,
Classicism in quantum physics,
Climate science,
Code, 38, 39, 45, 55, 62–64, 95, 101, 124, 155
Coded signal, 61–64
Cognition, 89, 161, 186, 188
Cognitive neuroscience, 6, 21, 23, 43, 62, 89, 126, 163, 168, 186
Coherence, 110, 144, 176, 177, 181, 189
Coherence theory of truth, 176
Collective welfare, 74
Commitment, ontological, 173, 175
Communication system, 61–64
Communication theory, 61
Composition of a system,
Computationalism, 164
Computer, 24, 36, 42, 61, 64, 111, 153, 156, 189
Computerese, 62
Comte, A., 8, 20, 25, 29, 115, 119, 121, 144, 160, 187
Concept, 6, 11, 12, 26, 27, 35, 37–39, 42, 61–64, 87, 89, 96, 97, 99, 105, 106, 108, 111, 116, 118–122, 125, 131–138, 143, 147–149, 154, 156, 167, 168, 170, 173, 174, 177, 178, 185, 186
Conflict, 9, 10, 19, 23, 25, 26, 28, 32, 54, 69, 84, 86, 88, 99
Conservation of energy, 131, 132, 136–138
Consistency, 12, 24–26, 188
Construct, 11, 82, 112, 129, 133, 154, 174, 176, 185, 187. *See also* Concept; Proposition; Theory
Construction, social, 9, 15, 89, 90, 125, 185
Constructivism,
Constructivism-relativism, 9
Constructivism, social, 185
Contingency. *See* Accident
Continuum of traders, 78
Convention, 133
Conventionalism, 20, 186
Copenhagen interpretation, 6, 127, 139, 141
Correlation, statistical, 72, 111, 116, 122, 147
Correspondence theory of truth, 45, 176
Counter-Enlightenment, 7, 32, 85, 185
Counterfactual, 123
Credence, 109–110
Credibility, 117, 118
Cremonini, C., 20
Crime, 38, 45, 72, 106, 111, 112, 164, 170
Crisis, 4, 30, 32, 33, 44, 61, 77–80, 145
Criterion, 3, 13, 145, 181–183
Cultural Revolution, 86
Cultural science, 42
Culture, 18, 19, 35, 39, 41, 45, 55, 69–72, 74, 82, 86, 87

D
Dalton, J., 6, 115
Daly, H.E., 74
Darwin, C., 6, 19, 28, 40, 55
Data, 18, 20, 21, 45, 58, 107, 109, 112, 115–117, 119–129, 146, 151, 153, 177, 186, 189

Davidson, R., 70
Davies, P., 61, 156
Debreu, G., 77, 78
De Broglie, L., 141
Deception, 153
Decoherence, 135, 142–144, 148, 152
Deduction, 186
De Finetti, B., 105, 116, 117
Dehaene, S., 161
De la Llosa, P., 16
Dennett, D.C., 155
Denotation. *See* Reference
Derrida, J., 4, 7, 8, 182
Descartes, R., 3, 17, 28, 120
Designation,
Determinin, 91, 108, 139–149
Determinism, 91, 108, 139–149
Dialectical logic, 86
Dialectical materialism, 9, 13, 19, 23, 83–86, 187
Dialectics, 10, 13, 23, 83–86
Díaz, P., 26
Dictatorship, 13, 69, 92, 97, 99, 100
Diener, B., 65
Diener, Ed., 65, 67
Dietzgen, J., 23
Digital metaphysics, 154
Digital physics, 6, 151–156
Dilthey, W., 8, 9, 29
Disequilibrium, 78, 80
Disorder, 38, 98–99, 102, 107, 108, 113, 144, 161
Distinguish ability, 167
Dogma, 5, 31, 72, 81, 88, 92, 108, 110, 124, 127
Douthwaite, R., 66
Dragonetti, C., 5, 24
Dualism, 3, 23, 42, 46, 86, 87, 131, 164, 185
Dubrovsky, D.I., 23
Dühring, E., 85
Dynamicism, 20–22
Dyzenhaus, D., 100

E

Eagleton, T., 88
Ebert, F., 100
Eccles, J., 23, 156
Economic
    equilibrium, 41, 77–81
    growth, 9, 59, 65–75
    imperialists, 28, 72
    theory, 27, 28, 30, 41, 69, 77–82, 90, 126
Economism, 32, 41, 71, 72, 86, 87, 91
Eddy, C.,
Egology, 8
Einstein, A., 6, 28, 40, 135, 141, 145, 154, 178
Einstein, Podolsky and Rosen (EPR), 144–146
Emergence, 17, 28, 42, 71, 80, 91, 134, 146, 185, 186
Emergency, 98–100
Emotion, 25, 63, 109, 160
Emotivism,
Empiricism
    classical, 187
    logical, 20, 187, 188(*see also* Logical; Positivism)
Energetism, 135
Energy, 6, 58, 59, 84, 116, 122, 123, 131–138, 140, 142–144, 148, 152–155, 185, 186, 188
Engels, F., 4, 8, 13, 83–93
Enlightenment, 7, 24, 32, 35, 36, 85, 92, 93, 131, 185, 188
Entanglement, 40, 144, 145, 149
Entropy, 142
Environment of a system, 189
Epidemiology, 8, 38, 106, 167
Epistemology, 20, 27, 30, 46, 81, 83, 88–90, 148, 155, 186–189
Equality, 27, 58, 72, 92, 101, 109
Equilibrism, 78
Euler, L., 6
Evans, J., 116
Event, 23, 52, 58, 63, 77, 106, 109–111, 120, 133, 139, 160, 174, 186
Everett III, H., 152, 153
Evidence, 5, 25, 28, 29, 45, 66, 106, 110, 112, 116, 117, 140, 153, 174, 186
Evo-devo, 21, 40
Evolution, 7, 22, 40, 84, 142, 155, 156, 163
Evolutionary psychology, 21, 24, 152
Existence
    conceptual, 23, 27, 42, 146, 174, 177, 186
    criterion,
    formal, 11, 174–177
    hypothesis, 27, 29, 129, 152, 153
    mathematical, 8, 10, 23, 27, 89, 141, 146, 173, 175, 177, 178
    predicate, 11, 174–176, 178
    real, 23, 25, 27, 81, 141, 146, 152, 153, 174, 175, 178, 186, 188
Existent, 9, 25, 31, 176, 186–188
Existential predicate, 174–175
Existential quantifier, 173, 175, 178, 186
Experiment, 4, 6, 17, 23, 29, 37, 70, 109, 120–122, 125, 128, 132, 140–143, 145, 146, 156, 176

Explanation
causal, 160
mechanismic, 159, 160, 162, 164, 186

F
Factual/formal distinction, 174
Fallibilism,
Falsificationism,
Farley, J., 74
Feller, W.,
Fertility, 44, 182–183
Feuerbach, L., 85, 90
Feyerabend, P.K., 6, 8, 18, 148, 151
Feynman diagram, 154
Feynman, R., 111, 132, 148
Fiction, 140, 146, 152, 174, 176, 186
Fictionism, 186
Financial sector, 79
Five Yeas Plan, 91, 92
Fodor, J., 6
Foster, N., 43
Fréchet, M.,
Free market, 70, 77, 78, 80–81
Frequentism,
Friedman, M., 30, 69, 78, 81
Fuller, L.L., 101
Function, 5, 23, 29, 42, 55, 66, 91, 122, 123, 128, 133–135, 143, 145, 159–162, 174, 186, 187
Functionalist anthropology, 41
Functionalist psychology, 159–164

G
Galbraith, J.K., 69
Galiana, I., 59
Galileo, G., 7, 13, 20–22, 27, 28, 31, 51, 55, 89, 115, 120, 125, 143, 178
GDP, 43, 58, 59, 67, 68, 70–75, 80, 123, 124, 126
Geisteswissenschaften, 5. *See also* Cultural science; Social science
Geneticism, 31, 32
Genetics, 13, 16, 22, 31, 40, 41, 55, 61, 63, 89, 106, 108, 116, 156, 160
Genome Project, 31
Gentile, G., 8
Geometrodynamics, 154
Gini index, 30, 70, 73, 123
Gintis, H.,
Giovanni, G., 8
God's existence, 174
Good, I.J., 116
Goodman, N., 27, 175
Govindarajan, V., 35
Graham, L.R., 16
Gravitation, 19, 135, 159
Gravitation theory, 128
Greene, B., 153
Greene, C., 59
Green Revolution, 44
Greenspan, A., 79, 80
Griffith-Jones, S., 77
Guessing, 107

H
Habermas, J., 8, 24, 29
Hamiltonian, 133, 134, 143
Happiness, 65–68, 73, 74
Harris, M., 41
Hartmann, N., 13
Hawking, S., 128, 142, 146
Hayek, F., 24, 30, 69, 81
HDI. *See* Human Development Index (HDI)
Hebb, D.O., 19, 163
Hebb rules, 42
Hegel, G., 3, 4, 6–8, 10, 13, 22–25, 29, 30, 83–87, 89, 90, 92, 101, 115, 185, 186
Heidegger, M., 4, 7–9, 26, 52, 176, 182, 185
Heisenberg's inequalities,
Heisenberg, W., 29, 127, 139, 148
Helmholtz, H.V., 131, 132
Henry, R.C., 139
Hermeneutics, 186, 188, 189
Hessen, B., 89
Hidden variable, 145, 146
High-level hypothesis, 129
Hindenburg, Paul von, 100
Historical materialism, 86–90
Hitler, A., 9, 100, 101
Hobbes, T., 101
Holbach, P.-H., 35, 36, 85
Holism, 8, 13, 38, 41, 42, 86, 87, 89, 90, 186
Housing problem, 35–37
Huarte de San Juan, J., 29
Hubel, D.,
Human Development Index (HDI), 43, 72, 73
Humanism, 24–29
Hume, D., 7, 8, 13, 20, 24, 88, 119, 121, 129, 144, 160, 162
Humphreys, P.,
Huppert, F., 65
Husserl, E., 4, 7, 8, 13, 29, 33, 112, 182
Hylorealism,

Hypothesis, 5, 25, 27, 29, 30, 32, 43, 67, 109, 110, 112, 116, 117, 121–123, 129, 134, 135, 145, 152, 153, 163, 164, 168, 185–187, 189
Hypothetico-deductive, 189

I
Idealism
  objective, 8, 12
  subjective, 8, 189
Idealism, philosophical
  objective, 186 (*see also* Platonism)
  subjective, 186 (*see also* Subjectivism)
Ideology, 13, 24, 80, 81, 88
Ig Nobel Prize, 25
Imagination, 4, 12, 30, 32, 151–152, 174
Indeterminacy, 108, 142–143, 146–148, 174
Indicators, 42, 57, 58, 66, 68, 71, 72, 74, 81, 106, 116, 117, 119, 121–129, 143, 146, 178
Individualism, 8, 30, 41, 42, 87, 186
Induction, 115–118, 186
Inductive logic, 110, 112, 115–118
Inductivism, 29, 115–117, 186
Industrial revolution, 66, 87
Inequality, 30, 42, 66, 67, 69–71, 92, 97, 110, 123, 146
Informatics, 61–64
Information
  code, 63, 64
  science, 42
  technology, 31, 42, 61, 64
Informationism, 31. *See also* Computationalism
Innate ideas, 163
Instrument, measuring, 5, 121–125, 127–129, 143
Integral development, 65–75
Interpretation, 6, 8, 11, 12, 64, 78, 88, 90, 105, 107, 108, 127, 139, 141–142, 144, 147, 152, 173, 178, 189
Intuition, 45, 117, 146, 163, 181, 186
Intuitionism, 8, 24, 186
Inverse problem, 116, 123
Irrationalism, 4, 8, 16, 24, 176, 188
Its from bits, 154, 156

J
Jeffreys, H., 105, 116
Jevons, S., 45, 78
Joule's experirment, 132
Jury theorem, 106
Justice, 11, 26, 38, 70, 74, 98, 101, 102, 112

K
Kahneman, D., 27, 68, 109
Kant, I., 4, 6–8, 20, 29, 33, 42, 84, 88, 112, 119–121, 142, 144, 162, 163, 186, 188
Kaya identity, 57–59
Keen, S., 81
Kelsen, H., 10, 38, 44, 100–102
Keynes, J.M., 19, 28, 30, 69, 80, 105
Khaldün, I., 41
Khomeini, A., 52
Kind, 4, 5, 11, 12, 18, 19, 27, 37, 39, 41, 45, 61–65, 67, 85, 95, 99, 106, 116, 119, 123, 125, 127, 128, 132, 133, 136, 137, 153, 162, 168, 169, 176–178, 185, 187
Klein, N., 72
Klingberg, T., 162
Knowledge
  pyramid, 167–171
  rosetta, 170
  system, 39, 44–45, 169
Koepsell, D., 53
Kranzberg, M.,
Kripke, S., 8, 11, 27
Kuhn, T.S., 6, 15, 18

L
Labriola, A., 90
Lakatos, I., 16, 30
Lalande, A., 24
Laplace, P.S., 108, 147, 148
Latour, B., 15, 53, 90, 125
Law, 10–12, 22, 31, 38, 42, 44, 45, 52, 54, 62, 84–86, 89, 95–102, 105, 106, 108, 115, 116, 121, 124, 125, 128, 133, 137, 139, 147, 148, 153, 155, 164, 170, 171, 175, 178, 185, 186
  statement, 116, 128, 186
Lawfulness, 96, 147, 148, 186
Learned ignorance, 108–109, 112
Lederbogen, F., 162
Legal positivism, 10, 38, 44, 101–102, 170
Legitimacy
  moral, 96, 97, 102
  political, 96–97, 102
Lenin, V.I., 8, 22, 23, 84, 85, 89, 91
Liberty, 5, 27, 72, 92, 96, 101, 109, 162

Life expectancy, 70–73
Likelihood, 105, 109–111
Lindley, D.V., 110, 116
Literacy, 18, 73, 87, 161
Localizationism, 42
Local realism, 145
Logical imperialism, 10, 27–28, 174, 176
Lomnitz, L., 36
Lora, E., 67
Ludd, J., 66

M
Mach, E., 8, 20, 29, 89, 115, 119, 121, 144, 160, 187
Machiavelli, N., 22
Mahner, M., 61, 141
Malinowski, B., 41
Mann, M., 99
Many-worlds
  interpretation of quantum mechanics, 152
  metaphysics, 11, 27
  semantics, 174
Mao, Z., 85
Marmot, M.G., 71, 162
Marxist philosophy, 13, 22, 30, 83–93
Marx, K., 8, 13, 23, 24, 28, 30, 69, 83, 85–93, 185
Material, 3–5, 9, 10, 19, 21–23, 31, 32, 35, 39, 54, 55, 57, 59, 62–64, 87, 108, 128, 131–135, 137, 139, 141, 147, 156, 159, 160, 169, 174, 178, 186–189
Materialism
  dialectical, 9, 13, 19, 23, 83–86
  emergentist, 187
  historical, 86–88, 90
  systemic, 19, 85, 164, 187
Materialism vulgar, 187
Mathematicism, 31
Mathematics, 8, 12, 17, 18, 28, 45, 46, 69, 83, 105, 151, 167, 169, 170, 173, 174, 176, 188
Mathematization, 31
Matrix
  philosophical, 15–33
  social, 16–17
Matter, 12, 16, 18, 23, 28–30, 40, 46, 53, 67, 84–89, 100, 102, 105, 106, 109, 110, 131–138, 141, 143, 145, 147–149, 151, 154, 155, 176, 185–189
Mauss, I., 67
McKeown, T., 38
McMullin, E., 151
Meaning, 11, 64, 118, 148, 175, 187–189
Measurements, 40, 107, 120–122, 125, 127, 128, 135, 140, 146, 174
Mechanics, 6, 7, 31, 84, 106, 116, 122, 126–128, 137, 144–148, 152, 160, 168, 177
Mechanism, 22, 31, 35, 39, 54, 55, 62, 92, 116, 155, 159–162, 164,186, 187
Menger, C., 45, 78
Mental experiment, 156
Mereology, 11
Merger, 4, 40–42, 188
Mermin, D., 139
Merton, R.K., 9, 26, 53
Merzenich, M.M., 162
Metaphilosophical norm, 183
Metaphysics. *See* Ontology
Methodological individualism, 8, 30
Mill, J.S., 69, 115, 144, 187
Mind-body dualism, 3, 42
Minsky, H., 79
Mirowski, P., 52, 81, 90, 181
Mlodinow, L., 142, 146
Modal metaphysics, 11, 27
Modernity, 5, 7, 20, 101, 183
Monism
  methodological, 181
  ontological (*see* Idealism; Materialism)
Monty Hall, 107–108
Morals, 21, 24, 26, 30, 36, 45, 54, 55, 68, 69, 90, 95–99, 101
Morgenstern, O., 80
Mosterín, J., 140
Multiverse, 151–153, 156
Mussolini, B., 8
Myrdal, G., 30, 70
Mysterianism, 8, 15–16

N
Nationalization, 126
Naturalism, 5, 12, 19, 20, 24, 187. *See also* Materialism
Nazism, 92, 99–101
Needham, J., 17
Negation, 25, 84, 85, 173
Neo-determinism,
Neo-Hegelianism, 8
Neo-Kantianism, 8
Neo-liberalism, 41
Neo-Platonism, 13
Neo-Thomism,
New Age, 136
Newton, I., 6, 7, 19, 25, 29, 55, 84–86, 89, 90, 108, 115, 128, 159

Nietzsche, F., 7–10, 24, 26, 66, 183, 185
Nominalism, 187

O

Objectivism. *See* Realism
Objectivity, 16, 29
Obscurity, conceptual, 85
Observable, 116, 121, 123–126
Observation, 4, 70, 121, 122, 127, 139, 140, 144, 146, 176
Observer, 6, 29, 51, 125, 139–141, 143–144
Omnès, R., 140
Ontology, 4, 19, 22, 25, 27, 30, 43, 46, 83, 85, 86, 88, 137, 148, 155, 160, 175–176, 178, 187, 189
Operationism, 120, 121
Opposite, 11, 23, 29, 43, 84, 136, 144, 169, 185, 187
Ordinary language, 105–107
Ortega y Gasset,
Ossification, 22–23, 89
Ostwald, W., 8, 115, 131, 135

P

Panlogicism, 10, 176
Paraconsistent logic, 25
Parallel universes, 151–156
Part-whole relation, 87
Pavlov, I., 23
Perception, 160, 163, 185, 187
Petroski, H., 52
Phenomenalism, 4, 7, 8, 10, 29, 119, 121, 129, 144, 187
Phenomenology, 3, 8, 9, 11, 182
Phenomenon, 109, 187
Philosophical pentagon, 26, 28–30
Philosophical system, 45–46, 83, 85
Philosophy
  evaluation criterion, 182, 183
  linguistics, 3, 8, 11, 12
Phobosophy, 3–13
Physicalism. *See* Materialism vulgar
Physics, 6, 8, 10, 11, 13, 19, 21, 22, 29, 31, 39, 40, 44, 52, 53, 61, 62, 81, 83, 86, 89, 116, 119–128, 131–134, 136–149, 151–156, 160, 163, 167–169, 173
Planck, M., 131, 141
Planning, 36, 67, 91–92
Plasticity,
Plato, 8, 12, 21, 22, 134, 174, 186
Platonism, 12
Pluralism, 101, 102, 181–183, 185, 187
Poincaré, H., 115, 116
Politics, 13, 16, 25, 27, 44–46, 66, 70, 72, 74, 87, 91, 95–96, 170
Popper, K.R., 6, 16, 23, 72, 115–117, 146
Positivism
  classical, 187
  logical, 7, 127, 187
Possibility
  conceptual, 27, 174
  real, 27, 153
Possible world
  metaphysics, 11, 27
  semantics,
Postmodern, 7, 15, 17, 25, 51, 58, 102, 183
Postmodernism, 4, 7, 8, 85, 90, 188
Pound, R., 44
Pragmatism, 8, 10, 90–91, 188
Praxiology, 39, 46, 53
Praxis, 10, 52–53, 90–91, 183
Predicate, 11, 116, 120, 137, 168, 174–176, 178, 188. *See also* Property
Presidentialism,
Pre-Socratics, 5, 12, 85, 140
Press, S.J., 110
Probability
  of causes, 108, 112
  conditional, 117
  inverse, 117
  objective, 106, 107, 109, 111
  prior, 109, 110, 112, 117
  of propositions, 105, 108, 117
  subjective (*see* Bayesianism)
Problematics, 37, 164
Progress, 6, 7, 9, 15–33, 38, 44, 66, 71, 73, 88, 101, 119, 183
Proof, 5, 25, 38, 155, 174–175, 177
Property, 4, 6, 25, 28, 31, 35, 40, 44, 54, 63, 84, 120, 122–124, 126, 128, 129, 131–137, 140, 142–144, 147, 156, 169, 173, 175, 182, 185–188
Proposition, 12, 15, 105, 108, 117, 160, 168, 170, 174, 177, 185–189
Pseudo-philosophy, 7, 8
Pseudoproblem, 11, 107, 142
Pseudo-science, 7, 26, 33, 72, 81, 90, 112, 153, 156, 171, 188
Psychological, 16, 24, 108, 112, 146, 156, 164
Psychology
  biological, 9, 21, 24, 160
  functionalist, 159–164
Psychoneural dualism, 23, 164
Psychoneural identity, 27, 29, 43
Ptolemy, 18, 20, 119
Putsch, 100

Q
Quanton, 135, 143–145, 148
Quantum theory, 6, 8, 13, 22, 40, 89, 108, 122, 127, 128, 135, 136, 139, 143–148
Quintanilla, M.A.,

R
Rand, A., 11, 79
Randomization,
Randomness, 106–108, 111, 112, 147. *See also* Chance
Ratioempiricism, 4
Rationalism, 4, 19–21, 24–27, 186, 188
Rationality, 4, 5, 12, 16, 24, 25, 27, 72, 79–80, 139, 148, 185, 188
Rationality, economic, 27, 79–80
Realism, 3–5, 19–22, 28, 29, 139–149, 162, 170, 182, 186–188
Reality, 3, 4, 11, 12, 16, 20, 21, 24–29, 31, 45, 51, 55, 57–59, 78, 80, 81, 83–93, 116, 125, 129, 131, 137, 144–147, 153, 159, 169, 170, 174, 176–178, 182
Reasoning ability, 162
Reduction, 69, 119, 120, 162, 186–188
Reductionism, 32, 40, 188
Reference, 6, 120, 122, 134, 135, 175, 178, 187
Refutability, 153
Relativism, 3, 9, 101, 102
Relativity, 22
Relativity theory, 6
Republic, 10, 91, 99, 100
Rescher, N., 181
Risk, 6, 46, 65, 67, 111, 164
Risk assessment, 110, 111
Robinson, M.B., 112
Romanticism, 3, 7, 88
Roubini, N., 81
Rousseau, J.-J., 3, 13, 38, 85
Rugg, M.D., 62
Rule of law, 95–102
Russell, B., 8, 12, 74, 115, 147, 173

S
Samuelson, P., 78, 80
Sarkar, C., 35
Sartre, J.-P., 4, 182
Savage, L.J., 105
Schlosshauer, M., 128, 142
Schmitt, C., 10, 100, 101
Schütz, H., 59
Science
- applied, 52, 53, 55
- basic, 43, 52–55, 61, 64, 154
- biosocial, 8, 42–43, 106, 167, 188
- cultural, 42
- factual, 22, 115, 170, 175, 176, 178, 188
- formal, 174, 175, 188
- natural, 8, 13, 17, 24, 29, 40–41, 46, 106, 155, 168, 188
- social, 3, 7, 8, 21, 30, 32, 38, 41–44, 46, 61, 62, 69, 72, 74, 88, 90, 93, 106, 120, 124, 133, 152, 160, 169, 170, 188

Scientific Revolution, 6, 7. 18–20, 88, 115, 121, 176
Scientism, 7, 21–28, 85, 185, 188
Sectoral approach, 18, 38, 39, 44, 45
Sectoral development, 71–73
Semanticmic,
Sen, A., 68, 69
Sextus Empiricus, 4, 16
Shannon, C., 61–63
Shirer, W.R., 99
Shorter, E., 76, 154, 164
Skepticism
- moderate, 22, 188
- radical, 8, 9, 15, 16, 188

Smelser, N.J., 82
Smith, A., 59, 65, 66, 69, 80
Sobel, N., 161
Social indicators, 66, 71, 125
Socialism, 23, 70, 83, 91–93
Socialization, 89, 91
Social justice, 70, 98
Social science, 3, 7, 8, 21, 30, 32, 38, 41–42, 44, 46, 61, 62, 69, 72, 74, 88, 90, 93, 106, 120, 124, 133, 152, 160, 169, 170
Socioeconomics, 19, 21, 30, 33, 42, 44, 52, 68, 171, 186
Sociologism,
Sociology of knowledge, 88–90
Someness, 173–174, 176, 186
Soros, G., 33, 69, 79
Soviet regime, 69
Spacetime, 6, 11, 135
Species, 84, 111, 132, 142, 147
Speculation, 3, 13, 24, 79, 81, 95, 115, 151, 153
Stanford Linear Accelerator, 154
States, 6, 10, 16, 17, 27, 30, 33, 38, 41, 53, 62, 67–73, 77–80, 87–89, 91–92, 95, 98, 99, 106–108, 111, 119, 122, 124, 127, 128, 134, 135, 137, 141–144, 152, 178, 186, 188, 189

Statization, 91
Stiglitz, J., 66, 72
Stockholm school, 70
Stove, D., 16
Strawson, G., 139
Stress, 67, 70, 123, 142, 153, 162
Strike, 98, 99
Structure of a system, 189
Subject, 10, 12, 16, 19, 20, 24, 29, 46, 61, 73, 105, 112, 120, 123, 125, 141, 142, 144–146, 153, 187, 189
Subjectivism, 13, 16, 29, 89, 112, 186, 189
Subjectivist interpretation, 8, 108, 141–142, 147
Submission, 153, 162
Supersystem, 21, 38, 43
Susskind, C.,
Swift, J.,
Swiss-Army penknife model, 43
Systemic approach, 5, 21, 35–41, 43, 44, 46
Systemic materialism, 19, 23, 164, 187
Systemics, 35–46, 189
Systems, 4, 5, 8, 10, 11, 18–21, 23, 26–28, 31, 35–46, 52, 57, 61–64, 70, 81, 83–85, 87, 88, 95, 101, 119, 120, 127, 143–145, 156, 159, 164, 169, 171, 175, 185–187, 189

T
Tarski, A., 167, 173, 176, 177
Tawney, R., 69
Technology, 3, 4, 10, 17, 18, 27, 31, 38, 42–44, 46, 51–55, 59, 61, 64, 69, 74, 87, 88, 93, 105–107, 131, 151, 156, 170, 176, 178, 189
Technoscience, 51–55
Tegmark, M., 152
Testability, 151, 153, 188, 189
Theory, 4–8, 11–13, 16, 20–22, 25–28, 30, 31, 33, 37–41, 44–46, 55, 61–64, 69, 77–84, 86, 89–91, 95, 108, 110, 111, 115–117, 119–129, 133, 135–137, 139, 141, 143–148, 152, 154, 156, 167, 168, 170, 175–178, 185–189
  confrontation with data, 86
Thesis XI, 90
Thomism, 9, 182
Thought experiment,
Tola, F., 24
Translational research, 54
Trial by numbers, 105, 111–112
Trigger, B.G., 30, 90
Truth
  criterion of, 3, 181, 182
  factual, 176–178, 189
  formal, 176, 177, 189
  partial, 15, 88, 137, 178, 189
  theories of, 45, 176
Turiel, E., 25

U
Uncertainty
  objective, 108
  relations (*see* Heisenberg's inequalities)
UN Charter, 101
Universal Declaration of Human Rights, 96, 97, 102
Universal science, 131, 169–170
Universe, 5, 6, 19, 21, 45, 121, 128, 135, 136, 139–142, 146, 147, 151–156, 173, 175, 186
Utopianism, 91

V
Van Quine, W.O., 167, 173, 175
Variance, 83, 146, 173, 188
Veda, 5, 139
Vendral, V., 155
Venn, J., 109, 168, 171
Verstehen, 8, 88, 189. *See also* Interpretation; Psychological
Victor, P., 72
Virtual
  particles, 153–154
  worlds, 153–154
von Helmholtz, H., 131
Von Wright, G.H., 116
Vos Savant, M., 107

W
Wallerstein, I., 41
Walras, L., 78
Wan, Poe Yu-tze, 42
Wealth, 17, 20, 65–75, 79, 95
Weber, M., 9, 18, 32, 159
Weimar Republic, 100
Welfare state, 30, 33, 70, 71, 80, 91
Well-being, 58, 65–75, 102
Wheeler, J.A., 27, 135, 152, 154
Whewell, W., 115
Whitehall study, 71
Wiesel, T.,

Wikström, P.-O., 112, 164
Wilczek, F.,
Wilkinson, R.G., 70
Wittgenstein, L., 6, 8, 9, 11
Wolfgang, P., 27, 139, 147

Z
Zeilinger, A., 128, 146
Zen Buddhism, 183
Zeno, 84
Zwerdlin, D., 67

Printed by Publishers' Graphics LLC